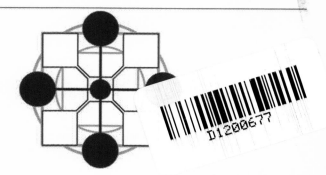

THE SCIENCE OF SCIENCES
AND
THE SCIENCE IN SCIENCES

VERSION 1.0

*THE SCIENTIFIC AND MATHEMATIC METHOD IN ANCIENT
AND TRADITIONAL AFRICAN CULTURE AND PHILOSOPHY*

WRITTEN BY

AFRICAN CREATION ENERGY

WWW.AFRICANCREATIONENERGY.COM

OCTOBER 10, 2010

AFTERWORD BY TEBNU AKHIR

AFRICAN CREATION ENERGY™
WWW.AFRICANCREATIONENERGY.COM

AFRICAN LIBERATION

SCIENCE, MATHEMATICS, AND TECHNOLOGY

(S.M.A.T.)

BOOK 1

ISBN 978-0-557-72871-8

Printed in the United States of America

AFRICAN CREATION ENERGY ALERT:

In Physics, Radiation is scientifically defined as any process by which the Energy emitted by one object, travels through a medium, and is eventually absorbed by another object. In Physics, a Black Body is scientifically defined as an ideal object that absorbs all electromagnetic radiation (Light Energy) that it receives, and when energized, a Blackbody is a perfect emitter of electromagnetic radiation. In Physics, a "Kerma" is an acronym for "Kinetic Energy Released in Material," and is the sum of the Radiant energy of the material divided by the mass of the material. The symbol above is a modification of the warning symbol for "Radioactive Material". In order to have Knowledge, one must first receive information, and in order to know if the information received is "Right Knowledge", one must have a method to properly process that information to determine if it is right or wrong. Only Right Knowledge or Correct information can be used, applied, and put into action as it is given. Information in any form must be radiated from one object to another object through some material; therefore Radiation and Energy are forms of Information. Information is the essence of Knowledge, and essential to life, therefore the proper processing of Information is the only way of life. This way of life was the way of life found in the oldest known African civilization called the Kingdom of Kerma. This book deals with the subjects of Science, Knowledge, Information, and the culture and way of life that is built upon the proper processing and utilization of Science, Knowledge and Information. The information and energy that is radiated and emitted by this book literally makes this book "Radioactive Material", and therefore, African Creation Energy's modified version of the "Radioactive Material" warning symbol appears here as a sign to signify the radiation of "knowledge" that is emitted from within these pages.

Proceeded if you Will

African Creation Energy

Creative Solution-Based Technical Consulting

Table of Contents

DEDICATION

The Science Of Sciences And The Science In Sciences

For The Separation And The Unification, The Liberation And The Salvation, The Solutions To The Problems Of The Dualities of Existence, Symbolized By The Two Lands, The Two Worlds, Which Are Inclusive Of:

The Living And The Dead, The Upper And The Lower, The Above And The Below, The Inside And The Outside, The Circle And The Square, The Sun And The Moon, The Day And The Night, The Infinite And The Finite, The Empirical And The Rational, The Unknown And The Known, The Abstract And The Concrete, The Introvert And The Extrovert, The Female And The Male, The Esoteric And The Exoteric, The Celestial And The Terrestrial, The Speculative And The Operative, The Subjective And The Objective, The Defensive And The Offensive, The Passive And The Aggressive, The Unconscious And The Conscious – And Many Other Binary

The Son Of The Sun
Daoud Of Ptah
Who Exists Forever

This Work From The Hut House Of Father Ptah Is Being Inscribed To Re-New The Great Work Of The Ancestors Which Had Become Consumed By "Parasites" To The Point Where It Could Not Be Completely Comprehended. This Work Is Being Renovated And Improved Upon For The Good Of The World In General, And The "Race" Known As African, Aethiopian, Black, Moor, Nubian, Or Nubun In Particular.

Let This Monument Last In The Hut House Of Father Ptah Throughout Existence As A Great Work Done By The Son Of The Sun, Daoud Of Ptah, So That He May Exist Eternally.

DEDICATION

HYMN TO PTAH

Hail to thee, thou who art great and old,
TA-TENEN, father of the Neteru gods,
the great god from the first primordial time
who fashioned mankind and made the gods,
who began creation in primordial times, first
one after whom everything that appeared
developed, he who made the sky NUT as
something that his heart SIA has created,
who raised it by the fact that SHU
supported it, who founded the earth GEB
through that which he himself had made, who
surrounded it with NUN and the sea, who
made the nether world and gratified the dead,
who causes RE to travel in order to
resuscitate them as lord of eternity
NU-HEH and lord of boundlessness lord of
life, he who lets the throat breathe and gives
air NEF to every nose, who with his food
keeps all Mankind alive, to whom lifetime,
limitation of time and evolution are
subordinate, through whose utterance HU
one lives, he who creates the offerings for all
the gods in his guise the great HAPY NUN,
lord of eternity, to whom boundlessness is
subordinate, breath of life for everyone who
conducts the king to his great seat in his
name, king of the Two Lands

Inspiration	from	My	Teachers
Revelation	from	My	Mind
Motivation	from	My	Brothers
Dedication	to	My	Family

1.0. THE PREFACE OF SCIENCE

What is the meaning of Life? What is the purpose for existence? Where did we come from? How did we get here? What happens after death? What is the origin of all existence? What is the ultimate fate of the Universe? These questions, as well as other problems, mysteries, curiosities and unknowns within and about Nature are the inquiries that have fascinated and captivated all inquisitive minds at some point in time throughout the ages. Historically, the study and investigation into any and all questions about Nature have been addressed in the various branches of Philosophy. The various answers derived from Philosophical inquiries gave rise to the many interpretations, explanations, and speculations that have been packaged into "philosophical technologies" called theologies, doctrines, dogmas, laws, morals, ethics, cosmology, and religions that use Philosophical ideologies as a base to try to provide answers and solutions to the inquisitive needs and probing minds of humanity. This **Natural Philosophy** that was responsible for **solving problems** was the predecessor and forerunner to modern **Science**. The word "Philosophy" means "the love of wisdom", and the etymology of the word "wisdom" is **"to see"** or **"to know"**, thus the word "Philosophy" also means "the love of knowledge" and is similar in meaning to the etymology of the word "Science" which is "**knowledge**". So in this regard, Philosophy is also "the love of Science".

Knowledge is acquired by the input or receiving of **Information** or **Data** to the Mind by way of the senses. Thus, Information is the substance that forms Knowledge, Philosophy, Science, and ultimately is the answer to all questions, **the key to eternal life**, and the foundation of everything that exists in Nature. "Information" is **the Natural Resource** that is needed by all beings, living or dead, and thus, Information is the essence of all existence and non-existence. Information is the foundation of Intelligence, Understanding, learning, survival, and life; for how would you know you exist without information? Information is Energy, and Energy is Matter, thus **Information is ALL**; both everything and nothing simultaneously.

The general scientific definition of Information is "that which **differentiates, separates,** or **distinguishes** one thing from another". The etymological root of the word "Science" is from the Latin word **"scire"** and the Proto-Indo-European word **"skei"** meaning "to separate, distinguish, cut, split, or divide". These etymologies and definitions are provided to further show the relationship in meaning between the words Information, Science, and Philosophy. Hence, an accurate method of processing Information is the **Path** to obtaining all that is needed.

As Natural Philosophy transformed into Modern Science, the explanations and answers in the form of doctrines and religions that were the result of Philosophical inquiries also transformed into modern Scientific Theories. Thus, it is only reasonable to expect that just as people accepted the doctrines, religions, and explanations that were the result of Philosophy without fully comprehending the initial question, problem, or method that led to the determination of a Philosophical conclusion, people are currently, and will likely continuously accept modern Scientific Theories without complete comprehension in the future. Therefore, the problems of people will persist as they become spellbound to the explanations provided to them by modern Scientist just as they were spellbound by explanations provided to them by Philosophers, Theologians, and Religious leaders in the past. This blind, unthinking, and zealous acceptance of scientific doctrine without proper comprehension was called **"Scientology"** in 1901 by a man named Allen Upward, and it was intended to be used as an insulting and derogatory term synonymous with ignorance. This term "Scientology" has been adopted in more recent times to identify a new-age "Religion of Science" or pseudo-Scientific based religion.

Organized Religion and Organized Science were at odds because they are two competing systems of providing the explanatory information to solve the problems in the lives of humanity. People accept Science because they can experience the explanations that are provided by Science, and Science works. However, the explanations provided by religion and various doctrines do not always work, which is why Science and Religion were at odds and are polar opposites. Thus, the transition from a "Religious" based way of life to a "Scientific" based way of life in the coming years is inevitable. Whereas Philosophy is the origin of Religious doctrines, the Philosophy of Science is the origin of scientific ideas and discoveries. The **Philosophy of Science** is "**The Science of Sciences and The Science in Sciences**", and this Philosophy of Science must be completely and fully comprehended and applied so that the future and coming scientific way of life is one that is truly based on knowledge, and not a contradictory Ignorant way of life of accepting information without first testing the information by way of "**The Science of Sciences and The Science in Sciences**".

The "**Science of Sciences and The Science in Sciences**" (which we will sometimes abbreviate as "the S.O.S. and S.I.S." or S.O.S.A.S.I.S.) is a way of life that is found throughout ancient and traditional African culture and philosophy, and thus, the coming and future scientific way of life should mark a return to traditional African culture and philosophy by people of African descent. With that brief introduction, we now can proceeded to discuss how this book entitled "The Science of Sciences and The Science in Sciences," which deals with The Scientific and Mathematic Method in Ancient and Traditional African Culture and Philosophy, came into being.

My family is one of the many families of the **African Diaspora** from African Ancestry, Background, and descent in the United States of America. Like most Black Americans or African-

Americans, either by way of the **Maafa** (African Holocaust), the Trans-Atlantic Slave trade, or some other event, most of the members of the African Diaspora (and even some people within Africa) have sparse, vague, or no knowledge and information at all about traditional African culture, philosophy, and way of life prior to slavery, colonization, exploitation, and other conflicts with foreign powers and ideologies. Thus, most Africans in the Diaspora and in Africa practice and accept cultures, traditions, philosophies, and a ways of life that are non-African. Ancient and traditional African culture and philosophy is very similar to "Scientific philosophy", and so many Africans in Africa and in the Diaspora are attracted to pseudo-Scientific based doctrines, perhaps, because these doctrines are similar to the traditional African way of life. There have been many movements amongst Africans in the Diaspora back towards traditional African culture. Some of these Black empowerment movements in the U.S.A. with pseudo-scientific ideologies and doctrines include **Noble Drew Ali's "Moorish Science Temple"**, the **"The Nation of Islam"** that was started by **Elijah Muhammad**, and **"The Nation of Gods and Earths** (also known as the **Five-Percent Nation**) started by **Clarence 13X**. Religiously, these various Black empowerment movements that had Scientific or pseudo-scientific undertones were also all **pseudo-Islamic**. Although these movements used various terms from "Islamic" religions, more than anything, they were movements to empower Africans in the Diaspora called the United States of America. And naturally, scientific minded Africans in the Diaspora would gravitate to these various "scientific sounding" pseudo-Islamic religious doctrines just as scientific minded individuals are naturally attracted to **"Sci-Fi"** movies and television shows. After several generations of my family members practicing the culture and religion of the very people who enslaved Africans, my Father and Mother took steps towards breaking away from those mental chains of bondage by converting to Islam. My Father joined The Nation of Islam, and after the death of Elijah Muhammad, my Father converted to what would be considered traditional **Sunni Islam**. During that point in time, my Father met my mother and started a family. Although the religion that we practiced at home was Islam, our family household was very much culturally **"Afro-centric,"** as we would regularly take part in various "African" cultural events. Although I was raised in the Islamic religion, it was always evident to me, even at an early age, that there were inherent, powerful, and more reasonable truths and explanations about life and existence within **Science** that was not provided by my religious beliefs.

I was raised as an orthodox Sunni Muslim for 17 years before my Father began to bring home books about a term called **"Nuwaupu"** written by an author named **Amunnub Reakh Ptah**. I have always been very curious and inquisitive by Nature since birth. I was a Natural Scientist, and my inquisitive mind always had questions that religion either could not answer, or did not answer sufficiently or reasonably to my satisfaction. Most of the books that my Father brought home about **"Nuwaupu"** dealt with questions and contradictions found within the three Monotheistic religions: **Judaism**, **Christianity**, and **Islam**. Other books dealt with topics such as various Conspiracies, U.F.O. phenomenon, Extraterrestrials, Time Travel, Egyptian History, Sumerian History, and a plethora of other scientific and pseudo-scientific subjects that, combined with very detailed artistic images of Black/African people, fascinated my inquisitive and creative mind. At first glance, it seemed to me that Nuwaupu was another pseudo-Scientific, pseudo-Islamic, Black empowerment "religion," but upon further investigation I soon discovered that there was much more to Nuwaupu than what **"meets the eye"**. Although I first read a book about Nuwaupu in **1998** A.D. and found the questions and contradictions about the religion of Islam raised by the author to be reasonable, I still practiced and considered myself a Muslim for the next **9 years** until the year 2007 A.D. During that time I obtained a Bachelors degree in Electrical Engineering with a minor in Physics, and I also obtained Master's degree in Mathematics. During that time, although I was still reading books about "Nuwaupu," I did not know what the word "Nuwaupu" meant. Also, during that time, my Father, who was very active in the **Nuwaupian Masonic Lodge and Shrine Temple** established by **Malachi York**, died. I think the death of my father, plus my own interest in the books about Nuwaupu, combined with the assistance that my Father's Nuwaupian Masonic and Shrine brothers provided to our family, motivated me to want to fully comprehend exactly **"what is Nuwaupu?"**

I began attending Nuwaupian African cultural functions, I learned the Nuwaupic language, I learned Nuwaupian prayers, meditations, and affirmations, and I regularly attended **Nuwaupian "Question and Answer Classes"** that were free for anyone to ask any question about any subject (science, pseudo-science, math, numerology, astrology, religion, philosophy, history, metaphysics, law, etc.), and substantial and insightful answers were regularly provided. Naturally this setting was a pleasure to someone as inquisitive as I. During my study of Nuwaupu, I was motivated to take part in the **African Ancestry DNA test** to determine exactly where in African my family came from, and also I made my first trip back to Africa in 2008 A.D. which marked the first time someone from my family had returned back to Africa since my ancestors were taken during the Trans-Atlantic Slave trade hundreds of years earlier. By studying Nuwaupu, I learned how my African Ancestral tribe migrated to West Africa from the area of the present day countries of Egypt, Sudan, and Ethiopia as part of the **Bassa** migration, and I was able to trace my lineage all the way back to the **Kushite/Napatan/Meroe** kings of **Egypt's 25th dynasty** and even further back to pre-dynastic Africans at the source of the Nile river in the area of the present day countries of Uganda, Congo, Kenya, and Tanzania. Studying Nuwaupu also motivated me to research **ancient and traditional African culture and philosophy**. It was through the study of "**Nuwaupu**" that I was able to go through the **doorway** to reconnect with who and what I truly was as an "**African**", both physically and mentally.

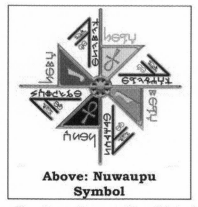

Above: Nuwaupu Symbol

While learning the "meaning" of "Nuwaupu", I first encountered the phrase **"The Science of Sciences and The Science in Sciences"** while reading and studying several different books written by **Amunnub Reakh Ptah**. In a book entitled "Bible Interpretation and Explanations, Booklet One", I encountered the term **"The Science of Sciences and The Science in Sciences"** where the writer **Amunubi Rahkaptah** provides a definition of "Nuwaupu" that states: *"NUWAUPU is pronounced NOO-WAH-POO. The meaning of the word Nuwaupu is **THE KNOWLEDGE, WISDOM, AND UNDERSTANDING**. Nuwaupu is ALL knowledge, ALL wisdom, ALL understanding – finite and infinite. **Nuwaupu is THE SCIENCE of sciences and THE SCIENCE in sciences**. Nuwaupu is the science of the ORIGINAL CREATIVE FORCES and the original spiritual science of the African Pygmies and Ethiopians in general".*

Also, in a book entitled "Wu-Nuwaupu", the writer Amunnub-Reakh-Ptah states: *"Nuwaupu is divided up into three degrees: Wu-Nupu, and Asu-Nupu and Naba-Nupu called Wu-Nuwaupu...It's also written Nuwaubu. Yet, here for Nuwaupians I will speak to it's pronounced as Noo-Wah-Poo or Nuu-Wuu-Puu. It stands for the **Right Knowledge, the Right Wisdom and the Right Overstanding**. The key to transform information into* *outformation.   Nuwaupu   is   all   comprehension...Finite   and infinite...*__*Wu-Nuwaupu is the science of sciences and the science in sciences*__*...Nuwaupu is the science of the original creative forces and the original forces of ether of the African people in general".*

Upon learning this definition for the word "Nuwaupu", my scientific mind was intrigued by the idea that this concept called "Nuwaupu", which in definition, sounded very much like **"The Scientific Method"** that I learned and had been taught about in school as the foundation upon which all operative Science is based, could have possibly been used not only as an operative Scientific method or **"Philosophy of Science"**, but also as a **"Spiritual Science"** or "way of life" in Ancient and Tradition African culture and philosophy. This notion set me out to do more research and search for evidence to confirm the existence of this **"Science of Science"** or "Scientific Method" in African Culture and Philosophy.

In doing further research, I came across the term **"The Science of Sciences"** in a book entitled "Introduction to The Nature of Nature" written by **Afroo Oonoo** where he states: *__Noone in its fullness is The Science of Sciences__ -- The Pioneer of Science. Science (knowledge) is THE PHILOSOPHY OF REASON (Page 114).   __NOONE is THE SCIENCE OF SCIENCES__ -- The Science of Sound Right Reason -- The Science of All Knowledge, All Understanding, and All wisdom (Page 236)."*

My research also led me to discover the term "**The Science of Science**" in the following sources:

- Manly P. Hall writes in his "The Secret Teachings of the ages that German Philosopher Johann Gottlieb Fichte defines Philosophy as "<u>The Science of Sciences</u>"

- In his 1933 book entitled "*Scientology, Science of the Constitution and Usefulness of Knowledge*", philosopher Anastasius Nordenholz used the term "Scientology" to mean "<u>science of science</u>"

- The term "<u>The Science of Science</u>" appears in a book entitled "*The Search for Truth on the Path of Reason*" by Alexi Osipov and also in a book entitled "*Philosophy of Science: A Very Short Introduction*" by Samir Okasha

- The term appears in the title to a paper entitled, "*A Mathematical Proof of the Definition of the <u>Science of Science</u>*" by Yi Lin and Yonghao Ma

- In a book entitled "*Real Science: What it is, and what it means*", the writer John Ziman uses the term "<u>The Science of Science</u>" to refer to Meta-scientist who are philosophers which study scientist to describe the workings and accomplishments of science.

My research about the term "The Science of Science" resulted in me discovering several papers and books about Philosophy as it relates to Science. Outside of the books written by Amunubi Rahkaptah and Afroo Oonoo, I discovered a book entitled "**The Science of Sciences**" published by Chinmaya Publications which talked about not only "Philosophical

Science", but also "Spiritual Science" or Science as a "way of life". This book entitled **"The Science of Sciences"** was a collection of essays about the concept of **"Brahma-Vidya"** which was called **"the Science of Sciences"** or **"the Science of God"** in Hindu culture and philosophy. Further research into this Hindu concept called "Brahma-Vidya" (which was called "The Science of Sciences" or "The Science of God") and the associated concepts and terminologies in the **Asian** Philosophical schools of **Hinduism**, **Buddhism**, **Jainism**, **Sikhism**, and **Taoism** led me to discover striking similarities and parallels between not only the operative "Scientific Method" (that is the Philosophical foundation of modern operative Sciences), but also to the Ancient and Traditional African Culture, Philosophy, and way of life that I had been motivated to learn about while studying **Nuwaupu**. Even the terminology used within Hindu Philosophy was similar to the wording and terminology I encountered in definitions of Nuwaupian Philosophy.

In Hindu philosophy, which is considered by some people as "The world's oldest living religion", as well as in the other Asian Philosophical schools mentioned, the Brahma-Vidya ("The Science of Sciences" or "The Science of God") is a practice or **method** (called **Yoga**) which leads to the **"end of suffering"**, liberation (called Moksha, **Zen**, **Nirvana**, or **Nibbana** depending on the tradition), and "knowledge of God". The **Buddha** (which means **"enlightened one"**, "Self-awakened one" or "**self created one**") developed a systematic method which he called

"**The Middle Path**" which was the practice of "**The Science of Sciences**" (Brahma-Vidya) and led to **liberation**, wisdom, enlightenment, heaven, and self-awakening. This method was called the "Middle Path" because the practitioner was expected to **moderate**, mediate, or **meditate** between the two extremes of indulgence and restriction, and **balance** and center one's self rather than swinging back-and-forth between the two extremes like a **plumb-bob pendulum**.

The first step of "**The Middle Path**" according to "the enlightened one" (called Buddha) in practicing "**The Science of Sciences**" was to understand "**The Four Noble Truths**". The Four Noble Truths are types of **Experiences**,

Above: Hindu Swastika

and each of the "Four Noble Truths" is symbolically represented by one of the arms of the **Hindu Swastika** symbol which also represent "**4 Rays of the sun**" and the "**4-main Cardinal directions**" (North, East, West, and South). If the arms of the swastika are facing **right**, it represented the **External** and if the arms of the swastika are facing **left**, it represented the **Internal**, and these two forms were the **Duality** of the Hindu deity **Brahma** who was said to have "**4 faces looking in 4 directions**". The swastika was said to be the "**heart seal of Buddha**". This swastika symbol influenced the "**Sun Cross**" or "**Sun Wheel**" used in **Gnosticism** (meaning **knowledge**) as well as "**The Black Sun**" symbol used in occult **Nazi** iconography which consisted of three swastikas rotated representing the **sun** at it's **rising**, **noon**, and **setting** positions. The Four Noble Truths (positions of the Sun or "experiences") as expressed in Asian philosophy are:

1. The Nature of Suffering (Problems, Questions, etc.)
2. The Origin of Suffering (Problems, Questions, etc.)
3. The Elimination of Suffering (Problem Solving)
4. The Path that Leads to the End of Suffering (The Method)

The fourth point of the "**Four Noble Truths**" (or **4 positions of the Sun** or "**4 types of experiences**") in Asian Philosophy is the actual Practice or Method (**Yoga**) that leads to **Liberation** (or **problem solving**) and is called "**The Noble Eightfold Path**" or "**The Wheel of Dharma**" which is used to gain **Truth** about **natural phenomena** and reality. The eight practices of the "Noble Eightfold Path" which lead to **Right Knowledge** (correct information) and **Right Liberation** (correct problem solving) according to Asian philosophy are:

Wisdom
1. Right View
2. Right Intention

Ethics
3. Right Speech
4. Right Action
5. Right Livelihood

Concentration
6. Right Effort
7. Right Mindfulness
8. Right Concentration

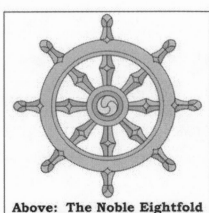

Above: The Noble Eightfold Path which leads to Right Knowledge and Right Liberation in Asian Philosophy

In the East Asian philosophy of **Taoism** (which means **"Path"** or "way"), concepts similar to those presented in the Wheel of Dharma of Hinduism and Buddhism are expressed symbolically in Taoism using the Taijitu (**Yin and Yang duality symbol**) surrounded by the **Ba-gua** or "**Eight tri-gram symbols**" which are said to represent the fundamental principles of reality in Taoism. The 8 tri-grams of the Ba-Gua are said to have came into being when "***the infinite created the finite***, *which created two forms Yin and Yang, which each divided into **4 forms** called lesser yin and greater yin (**Moon**), and lesser yang and great yang (**Sun**), which then divided into the 8 forms of the Ba-gua.*" The Ba-gua takes on two forms, a **Primordial** form and a **Manifested** form. The "Eight tri-grams" of the Ba-gua which are said to help **understand the dualities of Yin and Yang** in Taoism are:

- Qián - Heaven/Sky
- Duì - Lake/Marsh
- Lí - Fire
- Zhèn - Thunder
- Xùn - Wind
- Kan - Water
- Gèn - Mountain
- Kūn - Earth

Above: The Taoist Yin and Yang duality symbol surrounded by the Ba-Gua

The "four noble truths," "four positions of the sun," or "four types of experiences," can also be likened to the Buddhist concept of the **"4 Wise Monkeys of Nippon"** who are depicted with one covering it's eyes, one covering it's ears, one covering it's mouth, and one crossing it's arms who collectively represent **"see no evil, hear no evil, speak no evil, do no evil"** respectively. The type of experiences that the "4 wise monkeys" are supposed to represent include "looking the other way" or **"turning a blind eye"** or ignoring information, and thus being unable to do anything, and also "if one does not receive wrong information then one cannot do wrong actions".

Above: 4 Wise Monkeys of Nippon

The **"noble-eight fold path"** is also expressed in Zen Buddhism with the **Ensō circle** symbol which is likened to the **Ouroboros** symbol which depicts a serpent eating it's tail that is used in the **Gnostic**, **Hermetic**, **Masonic**, **Theosophical**, and **Alchemical** doctrines as a symbol for regeneration, continuity, self-reference, eternity, and the cyclical nature of the Universe.

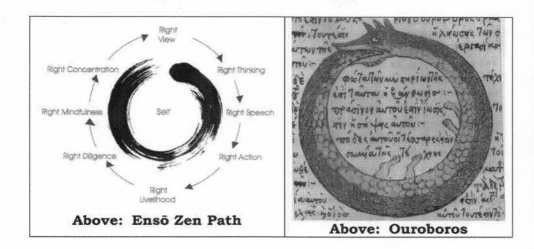

Above: Ensō Zen Path

Above: Ouroboros

These various Asian philosophies all have in common a set of **8** concepts (**Octo-gram** or "**squared circle**") that when combined with various "**practices**" or "**techniques**" as well as **transmitting**, chanting, or calling certain **tones** (names), the practitioner will be able to receive information on how to reconcile, "mediate" or "**meditate**" between opposing dualities (**solve problems**) to achieve some desired purpose or goal (liberation, enlightenment, or go to "heaven" and meet God, etc.). In Hinduism, the tone that is called on to meditate is called **Aum** (pronounced like Om, Ohm, or Ah-Um), and is considered the "**creative tone**" of the Hindu deity **Brahma**, whose name means "**to grow**," and who is considered a representation of infinite reality, all space, matter, energy, and time in Hindu religion. Thus, in these Asian philosophies, practicing "The Science of Science" (called Brahma-Vidya in Hinduism) leads to knowledge of reality (or knowledge of god called Brahma in Hinduism).

It is clear that the calling out of tones (Aum in Hinduism) called "chanting" as a way to "meditate" and obtain liberation (solve problems) and "know or communicate with god (or reality)" became "prayer" in the **Abrahamic** religions or way of life called Judaism, Christianity and Islam. In fact the word **Abraham** (which means **"to cross over"**), who is said to be "the **Father**" of the three major monotheistic religions, is very similar to the word **Brahma**. The **"eight-fold"** (octo-gram or octagon) path which was supposed to be the **"way"** to know reality (or god) became an Eight-sided building called the **"Dome of the Rock"** or **"Temple Mount"** which was the original **"Direction for Prayer"** (or place where one should direct their attention to receive answers to solve problems from reality or god) and is the one sacred site that is shared amongst all three Abrahamic religions: Judaism, Christianity and Islam.

In Arabic, the name for the **"Direction for prayer"** is called **Qiblah** or **Kiblah** or **Qiblih** and is related to the Hebrew term **Kabbalah** or **Qabala** which means **"receiving"** and is a

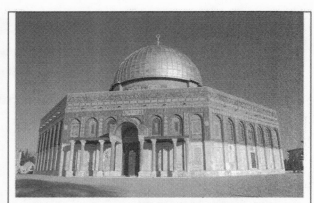

Above: The 8-Sided "Dome of the Rock" or Temple Mount Octagon sacred to Abrahamic religions

set of esoteric teachings intended to explain the mysteries of "the grower or Creator" (god) to mortals.

Above: The 8-Sided Octagon or "Squared Circle" ring used for Wresting and Martial Arts matches between two opponents

While the Abrahamic monotheistic religions applied an "esoteric" interpretation to the eight-principles of the "Science of Sciences" which is supposed to help mediate or meditate between two opposing forces, one of the exoteric applications of the "Science of Sciences" Octagon (eight fold path or squared circle) manifested in the world of **wresting** and **martial arts** where **two opponents "square off"**, try to **figure each other out** (problem solve), and try to defeat or conquer each other physically in an Octagon shaped ring.

It is apparent that the various Asian philosophies as well as certain practices in the Abrahamic monotheistic religions (Judaism, Christianity, and Islam) have been adopted from the Brahma-Vidya (called The Science of Sciences) philosophy of Hindu-ism or Sindu-ism which is considered by some people to be the world's oldest living religion. However, the question must be posed, **"How can Hindu-ism be the world's oldest living religion or "way of life" but Hindu people are not the world's oldest people?"** Moreover, Hindu-ism is said to be a mixture and collection of different ideologies and philosophies, and NO ONE is claimed as the founder of Hindu-ism. Because of my research into ancient and traditional African philosophy and culture that was motivated by my study of **Nuwaupu**, I could clearly see the African influences on Hindu-ism and the Brahma-Vidya concept called "The Science of Sciences". **Since it is a scientific fact that African people are indeed the world's oldest people, it would only be reasonable to search for the origins of the world's oldest way of life in Africa.** Also, modern scientific Genetic studies show that the Men (Haplogroup DE Y-DNA) and Women (Haplogroup L3 mtDNA) that became the various peoples of Asia were the descendants of East Africans who migrated out of Africa into Asia. So, it would also be reasonable to look for the origins of the various Asian philosophies, including "The Science of Sciences", in the place that is the origin of Asians – In Africa!

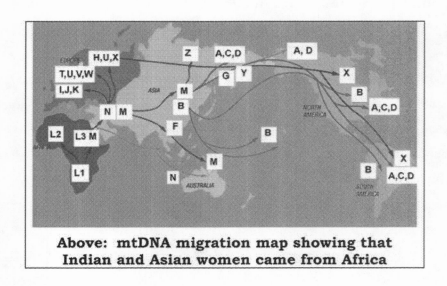

Above: mtDNA migration map showing that Indian and Asian women came from Africa

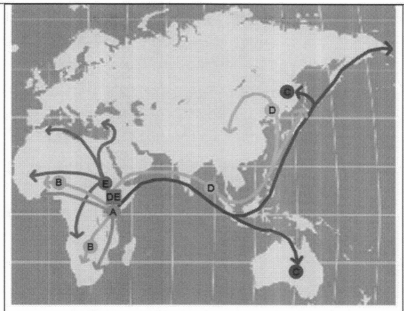

Above: Y-DNA migration map showing that Indian and Asian Men came from Africa

Modern Hindu-ism grew out of what is called the **"Historical Vedic religion"** which is a collection of "sacred writings" or philosophies with the oldest one called the **Rigveda** dating back as late as 1700 BC. An individual is considered an "Orthodox Hindu" if they accept

Above: Example of a Hindu Mandala

the Vedas as "Holy Scriptures." Prior to the "Eight-fold path" method of Buddha, the Rigveda provided methods in Hindu-ism to know "Brahma" (reality by way of The Science of

Sciences) as well as other "natural phenomenon" such as **Dyaus Pita** (the sky father) and other personifications of Earth, Water, Sun, and Wind using geometric figures called **Mandalas** which consist of a **square** with **four gates** containing a **circle** and a **central point**. The "Eight fold path" method (Yoga) or "Wheel of Dharma" that is the Method which leads to "The Science of Sciences" (Brahma-Vidya) was introduced by Buddha into Hinduism, and Buddha is dated to have lived as late as **600 BC**. So if the "Philosophy" or "Method" of the "Science of Sciences" can be found in Africa prior to 1700 BC or 600 BC respectively, then it is reasonable to conclude that "The Science of Sciences" is indeed of African origin.

The most notable empirical examples of "The Science of Sciences" that pre-date, and closely resemble the concepts that came later in Asian philosophy, are the concepts and stories associated with the African Deity **Ptah**. The similarities between the Hindu deity **Brahma** and the older African deity Ptah are overwhelming. The name **"Ptah"**, which means "to **Open**, **Separate**, or **Create**", is synonymous to **"Grow"** which is said to be the meaning of the name "Brahma". Also, Ptah is representative of All Space, Matter, Energy, and Time just like the Hindu deity Brahma. The tone, utterance, vibration, Breath, or pulsation that Ptah uses to create is called **HU**, whereas Brahma creates with the tone **Aum**. The breath or "vibration from the mouth" is called **Ātman**, "**the world soul**," in Hindu-ism is phonetically similar to the African deity **Atum** who is said to be **"the son of Ptah."** All of this evidence in addition to the fact that name **"Buddha"** is an Asian mispronunciation of the name **"Ptah"** having changed through time and tonal morphology from *"Ptah"* to *"Putah"* to *"Phutah"* to *"Pudah"* to *"Pubah"* to *"Buddha"* according to A.K. Coomaraswany in the book entitled *"The Origin of the Buddha Image & Elements of Buddhist Iconography"*. Moreover, the Hindu triad of deities Brahma, Shiva, and Vishnu who represent the concepts of Creation, Destruction, and Preservation share many characteristics and similarities to the African Memphite triad of deities Ptah, Sekhmet, and Nefer-Tum who represent the same principles respectively.

The philosophy, stories, concepts, and principles associated with the deity Ptah are called **"The Memphite Theology"** and the concepts presented in the story could be considered an African "theological" or "mythological" telling of **"The Science of Sciences."** Although the "Memphite Theology" is attributed to the Northern Egyptian city of **Memphis**, the text and details of the Memphite Theology come from an artifact dated to be from 700 BC called **"The Skabaka Stone"** attributed to the Southern Nubian Pharaoh **Shabaka Nefer-Ka-Re** from **Napata** (**Nu-Ptah**, meaning "People of Ptah"). Because the Pharaoh Shabaka established his capital at Memphis, and because the Ancient city of Memphis contained the **Temple of Ptah** and was considered the **"Het-Ka-Ptah"** meaning "House of the Spirit of Ptah," then the Shabaka stone text which described in detail the activities of the deity Ptah, naturally was associated with Memphis.

THE SHABAKA STONE MEMPHITE THEOLOGY OF PTAH

1. The Royal Titulary of Shabaka

The living Heru; Who prospers the Two Lands; The Two Ladies: Who prospers the Two Lands; The Golden Heru: Who prospers the Two Lands; King of Upper and Lower Kemet: Neferkare; The Son of Re: Shabaka, beloved of Ptah-South-of-his-Wall, who lives like Re forever.

2. The Restoration of the Text

This writing was copied out anew by his majesty in the house of his father Ptah-South-of-his-Wall, for his majesty found it to be a work of the ancestors which was worm-eaten, so that it could not be understood from the beginning to end. His majesty copied it anew so that it became better than it had been before, in order that his name might endure and his monument last in the House of his father Ptah-South-of-his-Wall throughout eternity, as a work done by the son of Re, Shabaka, for his father **Ptah-Tatenen**, so that he might live forever.

3. The Kingship of the God Ptah

— King of Upper and Lower Kemet is this Ptah, who is called the great name: Tatenen South-of-his-Wall, Lord of eternity —.

4. — The joiner of Upper and Lower Kemet is he, this unifier who arose as king of Upper Kemet and arose as king of Lower Kemet.

5 [—Intentionally Left Empty—]

6 —— "Ptah who is self-begotten, and gave birth to Atum: "who created the Nine Neteru."

7. The Unity of Horus and Ptah

Geb, lord of the gods, commanded that the Nine Neteru gather to him. He judged between Heru and Set; he ended their quarrel. He made Set the king of Upper Kemet in the land of Upper Kemet, up to the place in which he was born, which is **Su**.

8. And Geb made Heru King of Lower Kemet in the land of Lower Kemet, up to the place in which his father was drowned (namely **Pa-Seshat-Tawi**) which is **"the Division-of-the-Two-Lands."**

9. Thus Heru stood over one region, and Seth stood over one region. They made peace over the Two Lands at Ayan. That was how the division of the Two Lands came about.

10. Geb says to Set: "Go to the place in which you were born."
Set says: Upper Kemet.

11. Geb says to Heru: "Go to the place in which your father was drowned."
Heru says: Lower Kemet.

12. Geb says to Heru and Set: I have judged between you **"I have separated you."** — Lower and Upper Kemet. But then it seemed wrong to Geb that the portion of Heru was like the portion of Seth. So Geb gave Heru his inheritance, for he is the son of his firstborn son.

13. Geb says to the Enneanad (the Nine Neteru, Gods): **"I have appointed Heru, the firstborn.**

14. Him alone, Heru, the inheritance.

15. To his heir, Heru, my inheritance.

16. To the son of my son, Heru, the Jackal of Upper Kemet —

17. The firstborn, Heru, the **Opener-of-the-ways**.

18. The son who was born — Heru, on the Birthday of the Opener-of-the-ways." Then Heru stood over the land. He is the unifier of this land, proclaimed in the great name: Ta-tenen, South-of-his-Wall, Lord of Eternity. Then sprouted the two Great Magicians upon his head. He is Heru who arose as king of Upper and Lower Kemet, who **united the Two Lands** in the Nome of the Wall, the place in which the Two Lands were united. Reed and papyrus were placed on the double door of the House of Ptah. That means Heru and Set, made peace and united. They **fraternized** so as to **cease quarrelling** in whatever place they might be, being united in the House of Ptah, the **"Balance of the Two Lands"** in which Upper and Lower Kemet had been judged.

19. This is the land ------the burial of <u>Ausar</u> in the <u>House of Sokar</u>. ------ Auset and <u>Nebt-het</u> responded quickly without delay, for Ausar had **drowned in his water**. <u>Auset</u> and <u>Nebt-het</u> looked out, saw him and attended to him.

20. Heru speaks to <u>Auset</u> and <u>nephthys</u>: "Hurry, grasp him ---."

21. <u>Auset</u> and <u>Nebt-het</u> speak to Ausar: "We come, we take you ---." ------ They acted quickly and brought him to land. He entered the hidden portals in the glory of the lords of eternity. ------.

22. Thus Ausar came into the earth at the royal fortress, to the north of the land to which he had come. And his son Heru arose as king of Upper Kemet, arose as king of Lower Kemet, in the embrace of his father Ausar and of the gods in front of him and behind him.

23. There was built the royal fortress at the command of Geb ---.

24. Geb says to Tehuti: ------

25. Geb says to Tehuti: ------. ------.

26.

27. Geb says to Auset: ------

28. Auset tells Heru and Seth to come.

29. Auset says to Heru and Seth: "Come ------.

30. Auset says to Heru and Seth: "Make peace ------.

31. Auset says to Heru and Seth: "Life will be pleasant for you when ------.

32. Auset says to Heru and Seth: "It is he who dries your tears ------. ------

33. Auset says to ------

34. Auset says to ------

35. Auset says to ------

36-47: ---DAMAGED--- THIS IS THE SECTION ON UNIFICATION

48. The Gods who came into being in Ptah:

49. Ptah-on-the-great-throne ------ Ptah [Horus?]------ who bore the gods.

50. Ptah-Nun, the father who made Atum.
    Ptah [Thoth?] ----------- who bore the gods.

51. Ptah-Naunet, the mother who bore Atum. Ptah -----------

52. Ptah-the-Great is heart and tongue of the Nine Gods.
    Ptah ----------- Nefertem at the nose of Re every day.

The Theology of Memphis

53. **(logoism)** There took shape in the **heart of Ptah (Sia)**, there took shape on the **tongue of Ptah (HU)** the form of **Atum**. For the very great one is Ptah, who gave life to all the gods and their ka (Spirit) through this **heart** and through this **tongue**, in which **Heru** had taken shape as Ptah, in which **Tehuti** had taken shape as Ptah.

54. Thus **heart** and **tongue** rule over all the limbs in accordance with the teaching that it (the **heart**, or he: Ptah) is in every body and it (the **tongue**, or he: Ptah) is in every mouth of all gods, all men, all cattle, all creeping things, whatever lives, thinking whatever it (or he: Ptah) wishes and commanding whatever it (or he: Ptah) wishes.

55. Ptah's Ennead is before him as teeth and lips. They are the semen and the hands of Atum. For the Ennead of Atum came into being through his semen and his fingers. But the Ennead is the teeth and the lips in this mouth which pronounced the name of every thing, from which Shu and Tefnut came forth, and which gave birth to the Ennead.

56. Sight, hearing, breathing—they report to the heart, and it makes every understanding come forth. As to the tongue, it repeats what the heart has devised.

57. **(natural philosophy)** Thus all the gods were born and his Ennead was completed. For every word of the god came about through what the heart devised and the tongue commanded. Thus all the faculties were made and all the qualities determined, they that make all foods and all provisions, through this word, to him who does what is loved, to him who does what is hated. Thus life is given to the peaceful and death is given to the criminal.

pan-en-theism

58. Thus all labor, all crafts are made, the action of the hands, the motion of the legs, the movements of all the limbs, according to this command which is devised by the heart and comes forth on the tongue and creates the performance of every thing.

59. Thus it is said of Ptah: "He who made all and created the gods." And he is **Ta-tenen**, who gave birth to the gods, and from whom everything came forth, foods, provisions, divine offerings, all good things. Thus is recognized and understood that he is the mightiest of the gods. Thus Ptah was satisfied after he had made all things and all divine words.

60. He gave birth to the gods, He made the towns, He established the nomes, He placed the gods in their shrines, He settled their offerings, He established their shrines, He made their bodies according to their wishes. Thus the gods entered into their bodies, Of every wood, every stone, every clay, Every thing that grows upon him In which they came to be.

The royal residence : Memphis is the city of Ptah-Tenen

61. Thus were gathered to him all the gods and their ka (Spirit), Content, united with the **Lord of the Two Lands**.

62. The Great Throne that gives joy to the heart of the gods in the House of Ptah is the granary of Ta-tenen, the mistress of all life, through which the sustenance of the Two Lands is Provided, owing to the fact that Ausar was drowned in his water. Auset and Nebt-het looked out, beheld him, and attended to him. Heru quickly commanded Auset and Nebt-het to grasp Ausar and prevent his drowning (i.e., submerging).

63. They heeded in time and brought him to land. He entered the hidden portals in the glory of the lords of eternity, in the steps of him who rises in the horizon, on the ways of Re the Great Throne.

64. He entered the palace and joined the gods of Ta-tenen Ptah, lord of years. Thus Ausar came into the earth at the Royal Fortress, to the north of the land to which he had come. his son Heru arose as king of Upper Kemet, arose as king of Lower Kemet, in the embrace of his father Ausar and of the gods in front of him and behind him.

Although the physical artifact of the Shabaka Stone dates back to 700 BC, the wording and text contained on the Shabaka stone is much older. The earliest dates for the text of the Shabaka stone place it as early as 3200 BC with the first Pharaoh of the **First Dynasty** who **united Upper and Lower Egypt** named **Menes** (or **Narmer**). Menes, whose name is derived from the African **fertility** and **creation** deity named **Min**, established his first city at **Min-nefer** which eventfully became called "Memphis". It is suggested that the

Above: Shabaka Neferkare, inscriber of the "Shabaka Stone" Memphite Theology

Above: "The Shabaka Stone", a 2ft by 4ft slab describing the "Science of Sciences" as it relates to the African deity Ptah

ideas and concepts presented in the "Memphite Theology" of the Shabaka Stone were used by Menes as the "**Spiritual Science**" and philosophical doctrine of Menes' newly created city of Memphis (Min-nefer) during the First Dynasty. The City of Memphis was named after Menes and the African Creation deity Min, and Memphis contained the "Temple of Ptah," which considering Ptah's function, would more appropriately be called the "**Laboratory of Ptah**" or "**Ptah's Science Temple**."

Above: Menes or Narmer, founder of the 1st Dynasty of Egypt and the City Memphis

The city of Memphis was an Ancient African Industrial city that existed for the purpose of developing, growing, and creating **Science** and **Technology**. The African creation deity named Ptah, who was associated with growing, developing, creating, craftsmen, and artisans naturally had a "Philosophy of Science" in Memphis in order for the city to be the cult center of Science and Technology. Thus, the city of Memphis, which was the center for Ptah, gave rise to operative Scientist, Architects, Engineers, Physicians, and Builders, as well as "Social Engineers" or Philosophers. Some of the earliest operative expressions of the "**Memphite Theology of Ptah**" (**The Science of Sciences**) came during the 3rd dynasty of Egypt (around 2600 BC) from the vizier of Pharaoh **Djoser** named **I-M-Hotep** who later was deified as a god and "**son of Ptah**." Imhotep is considered the world's first architect, engineer, and physician. Imhotep is credited with having designed the first pyramid (The Step Pyramid of Djoser), as well as inscribing the first **operative empirical "Scientific Method"** on the world's oldest surviving Medical text (called "The Edwin Smith Papyrus") where he **systematically outlines the Scientific**

Methodological procedure of 1) **examination**, 2) **diagnosis**, 3) **treatment** and 4) **prognosis** as a means to **solve problems, heal,** or **Liberate the suffering** of 48 traumatic injury cases. This Scientific Method of problem solving was applied in a "rational" sense by the 18th Dynasty (around 1500 BC) Scribe **Ahmose son of Iy-Ptah** in a Mathematical papyrus where he states that *"Mathematics (the mental process of reasoning) is the method to inquire into Nature and obtain **Right Knowledge** of all Mysteries and Unknowns."*

| Above: I-M-Hotep son of Ptah, author of the world's first known empirical "Scientific Method" | Above: I-M-Hotep's Medical Papyrus using the Operative Scientific Method as a means of Liberating ailments |

Some of the earliest social expressions and applications of the **"Memphite Theology of Ptah"** (**The Science of Sciences**) came from **Ptah-Hotep**; the vizier of Pharaoh Djed-Ka-Re around 2500 BC during Egypt's 5th Dynasty. Ptah-Hotep is considered "the world's first Philosopher" and is credited with writing "**The Maxims of Ptahhotep**" which is called by some people "The Oldest Book in the World." Examining the philosophy of Ptah-Hotep, it is clear to see how the principles and ideas presented can be applied in both the operative sense of Science, Technology, and Development, as well as in the social sense as a "method" or "way of life." Some of the selected passages of Ptah-Hotep's philosophy are below:

Selected Maxims of Ptah-Hotep

* Great is the **Truth**
* **Right Action** can be measured with a **plumb-line**
* **Un-truth** and **Wrong Action Fails** in the long run
* Stop wrong Action Immediately in order to find Truth
* You cannot perform wrong acts if you truly know Reality
* Through your life, act with Reason and Moderation
* Listening benefits the Listener

Above: Ptah-Hotep, considered "The world's first Philosopher and Social Engineer" and author of "The Maxims of Ptahhotep"

* Fully experiencing an event leads to comprehension
* Ignorance accomplishes nothing
* Transmit only after you have learned and received
* Do not blame or criticize people who are not creative, and do not boast of your own creativity
* Be wiling to learn from any source, and do not let your knowledge swell your Ego
* Do not repeat information that has not been verified

The "Memphite Theology" describes Ptah's method of creating by "imagining creation in his Heart (**Sia**) and Speaking the creation with his tongue (**Hu**)." This "method" that Ptah uses to create is the Ancient African version of the modern "**Scientific Method**" or "**Science of Sciences**" because in order for Ptah to create, he forms a "**Hypothesis**" or **belief** (imagination) in heart (**Sia**) that he can create, then he carries out the act of creating or "**Experimenting**" by speaking the Hypothesis into existence with his tongue (**Hu**). After the hypothesis of the heart Sia, has been tested with the tongue Hu, the creation exists and can be **observed** and **experienced**, and the cycle of creation and science continues. The Judeo-Christian Religious world

Above: African Creation Deity PTAH

expressed this concept in the chapter called "Genesis" in the Bible where god creates something, then has to "**see that it is good**," much like a **Scientist testing a hypothesis** and seeing if the result of the **experiment** is "good" (i.e. if the Hypothesis is correct). This is why **"The Science of Sciences"** is also called **"The Science of God,"** because it is the method that God uses to create, thus it is also **"The Science of Creating"** and **"The Science of Creation."** Philosophically, the concepts of the Heart of Ptah called **Sia** or **Saa**, and the tongue of Ptah called **Hu**, are comparable to the concepts in Greek Philosophy called the **Nous**, representing the **"World Soul"** or **"The Mind"** and the **Logos**, representing **"The Word"** which Greek philosophers attributed to the source of **Creation**. The "Memphite Theology" describes Ptah as "the center of the Universe," and perhaps not coincidentally, the center of the "Shabaka Stone" has been damaged by someone cutting a square in the center and then scratching lines circularly radiating from the square which damaged significant portions of the text of the "Memphite Theology" especially the section that deals with Ptah as the Unifier of Upper and Lower Egypt and center of the Universe. Although archeologists say that the partial destruction of the text of the Shabaka Stone was done accidentally by someone "not knowing what it was," the pattern of destruction appears intentional and deliberate. However, despite the destruction, what is significant as it relates to "The Science of Sciences and The Science in Sciences" is that the text describes Ptah as the **"Divider of the Two Lands,"** **"The Unifier of the Two lands,"** and the **"Balance of the two Lands, the two ladies."** These titles attributed to Ptah make him an African deity of Science as a **divider**, **separator** (Opener), or **Analyst**, and a **builder**, **unifier** or **Synthesizer** of Scientific Natural Phenomenon. In order to comprehend how this is significant to "The Science of Sciences and The Science in Sciences," we must first discuss what "the Two Lands" and "The Two Ladies" represent in Egyptian mythology.

DUALITY: The Two Ladies, symbolic of The Two Lands

Nekhbet: Upper Egypt — Wadjet: Lower Egypt

In Egyptian Mythology, "**The Two Ladies**" are **Wadjet** and **Nekhbet**. Wadjet was depicted as a serpent or Egyptian cobra, and Nekhbet was depicted as a vulture. Wadjet represented one land, "Lower Egypt (Northern Egypt)" and Nekhbet represented the other land, "Upper Egypt (Southern Egypt)." While the act of "unifying the two lands" (or "the two ladies") was a real physical political act, it also held symbolic significance as a "unification" or coming together of opposite dualities: the aggressive and the passive, the mental and the physical, the sun and the moon, etc. In Science, these dualities are the Observer, and that which is being Observed, or the One Experiencing and the One Providing the experience, or simply the External World of Nature and the Internal World of The Mind. This duality of "the Two Lands" or "The two Ladies" manifested in early modern Science as philosophical debates between the concept of **Empiricism** supported by European philosophers Aristotle, John Locke, David Hume, and Francis Bacon, and the concepts of **Rationalism** supported by European philosophers Socrates, René Descartes, Immanuel Kant, and Baruch Spinoza. The duality of Empiricism and Rationalism, *"Nature vs. Nurture,"* or the Physical World and the Abstract Mental world took on various symbolic representations in Egypt and throughout Africa which included **Life** and **Death**, the Over-world and the Underworld, the **Sun** and the **Moon**, and the Egyptian philosophical schools of **Heliopolis** and **Hermopolis** that venerated the deities **RE** and **Tehuti** respectively. While these concepts of **Experience** and **Reason** were polarized in the philosophy of Heliopolis and Hermopolis, they were unified in the philosophy of Memphis (The Science of Sciences and The Science in Sciences) with the principle of **PTAH** at the center. Whereas **RE** represented the Sun, life, and Empiricism (external experiences); **Tehuti** represented the Moon, the Underworld (Death), and Rationalism (mental experiences).

| RE: Empirical | TEHUTI: Rational |

One of Tehuti's symbols was the **Baboon Monkey**. In Egyptian mythology, **four baboons** named Babi, Astennu, Hapi, and Aani (the god of **Equilibrium**) guarded the "**Lake of Fire**" in the Egyptian underworld (Rationalism, mind).

In some Egyptian texts there were **8 baboons** in the **solar bark of Re**. The baboon had connections to both the sun and the moon because the baboon is **Diurnal** (active during the day and night) and also baboons would also **bark** at the

Above: 4 Baboon Monkeys Guarding the "Lake of Fire"

rising sun. Symbolically, successfully passing through the "Lake of Fire" in the underworld would lead to seeing the "**sunrise**." The "Lake of Fire" is where modern religions get the concept of "**Hell Fire**." The "**Fire-Water**" or "**flammable liquid**" combined the symbolism associated with Fire and Water. In terms of the esoteric Alchemical symbolism associated with "**information**", "**fire**" was symbolic of "**information**" or "**knowledge**" (the sun) and "**water**" was symbolic of "**mythology**" or "**symbolism**" (mathematics and the moon) that enables comprehension and use of the information (wisdom). Thus "**Fire-Water**," "**a lake of fire**," or "**flammable liquid**" would be a very potent and powerful form of mythology or symbolism (water) that could be **ignited** (used) when combined with the right information (fire). This "Fire-Water" was the symbolism and mythology (mathematic rationalism) associated with Logic and Reason requisite for "Problem Solving" (Liberation) because only reasonable thoughts and symbolism can ever be used and applied, whereas unreasonable thoughts, symbols, and hypotheses fail when put to the scientific test of experimentation. Just as the "Science of Sciences" method of problem solving, obtaining information, and acquiring knowledge of Nature was symbolized in Religion and Mythology as "**Getting to know God**," likewise the inability to solve problems and misinformation (lies) were symbolized in Religion and Mythology as the **Devil**. In Hebrew, the meaning of the word "**Satan**", which is another name for "**The Devil**" is "adversary or obstacle"; thus the Devil or Satan is a religious mythological symbol for "**a Problem**." So in the Christian Bible when Jesus says "**Get behind me Satan**" it is a statement of problem solving, liberation, or putting one's problems behind them. Also, when Christian people make the statement "**The Devil is a Liar**" they are unknowingly and symbolically saying that lies or misinformation is a problem that must be solved. It is said in religious mythology that the Devil "stands in the **Lake** of fire called Hell" wielding a **trident** or **pitch fork**. The Abyss, "chaotic water," or "lake of fire" that became known as Hell in modern religious mythology was deified as the Mesopotamian deity named **Apsu** also called **Ensu** (Ensu is also related to **Ensō**, the Zen path), **Suen**, and **Sin**. Apsu or "Sin" was considered a "**Lord of Wisdom**" and **Lunar deity** (related to rationalism) associated with "**chaotic water**." According to the book entitled "*A Manual of Sumerian Grammar* *and Texts*" by John Hayes, the name of the deity Apsu in Mesopotamia transformed into Apa-Su-Eden and eventually into the name of Greek deity **Poseidon**. Poseidon was a Greek deity of **water** and also wielded the **trident** staff. The trident staff of Poseidon actually has four points symbolic of the "Science of Sciences" method where the 3 points at the top are the 3 day positions of the sun, and the fourth point is the long handle symbolic of night, the underworld, or the chaotic waters ruled by Poseidon, Apsu, Sin, and "the Devil" in religion and mythology. Variations of the trident symbol are also used as the Alchemical symbols for the radioactive elements named **Neptunium** (♆) which is a by-product of **Nuclear Reactors**, and **Plutonium** (♇) which is used to make **Nuclear Bombs**. The atomic numbers of Neptunium and Plutonium are **93** and **94**, and the Periodic Table symbols are **Np** and **Pu** respectively. The trident of Poseidon is geometrically similar to the Greek letter Psi (Ψ). The Greek letter Psi (Ψ) is phonetically identical to the "Sci" in the word the word "Science". The tri-dent or Quad-dent symbolized by "Psi" or "Sci" at the beginning of the word Science (Ψ-ence) is an indication of the "Four movements of the sun" or "The Science of Sciences" as the "beginning" or starting point of Science. The trident or letter "Psi" is also called a **Sai** and is used as a dagger weapon in **Nippon Martial Arts**. In operative Science, the letter Psi is used as a symbol related to water or waves in a variety of fields. The letter Psi (Ψ) is used as a symbol for "the **Potential energy of water**", the "**Steam Function**" in **Fluid Mechanics**, and the "**Wave**

Function" of particles in Quantum Physics. "Psi" is also a symbol for **Psychology**, **Psychiatry**, and the beginning of the word "**Cyclops**" (Psi-clops), symbolic of the "**third eye**" of the mind. Poseidon, whose named derived from the Mesopotamian "lord of Wisdom" called Apsu or Sin, is also related to the Roman deity of water called Neptune whose name is derived from the Mesopotamian "lord of Wisdom" called **Nebo**. The deity of "chaotic water" in Ancient Egypt was called "**Nun**", and **Nuwn** is also the name for the "**sea monster**" or "**whale**" in the Judeo-Christian Bible. In Modern Hebrew, the word for "whale" is **Leviathan**, which literally means "the law of Sin".

Again, the Mesopotamian deity named **Sin** is a moon deity associated with water and thus rationalism, therefore the "Law of Sin" or Leviathan is actually the "**laws of the mind**" called **logic** and **reason**. Leviathan, the Devil, and Hell have such negative connotations in religious mythology because if an individual is unable to properly use the laws of the mind

Baphomet
separate and unite
symbolic of "Problem Solving"

called logic and reason, then their life can indeed become very troublesome, problematic, and a living "Hell." The need to use logic and reason to "problem solve" or "conquer the Devil" is necessary for life, and this is why the words **"SOLVE"** (separate) and **"COAGULA"** (unite) appear on the right and left arms respectively in the iconography of another symbol of "Satan" named **Baphomet**. Not following the laws of the mind and being illogical and unreasonable will lead to unfavorable circumstances, and in religious mythology the individual is symbolically considered **"consumed by Nuwn"** or **"swallowed by the Whale"** (as in the Jonah or **Yaanus** story) or **burned (Cremated)** and drowned in the "**Lake of Fire**." To be illogical and unreasonable is the real meaning of "**committing a Sin**," and Sin historically was the Mesopotamian god of the Moon, Water, Wisdom, and Rationalism. The modern religious concept of "Sin" is absurd because it contradicts the religious concept of the "all powerful God" by implying the notion that an individual can perform an action that the all powerful creator did not want them to perform. The religious concept called Sin is itself a logical fallacy. In reality, anything that you should not do, you cannot do, and do not have the ability to do. Thus, the true almighty Grand Architect of the Universe designed the Universe so that any undesirable action could not possibly be performed, and therefore any and everything that is desired or willed to be done within the universe, can be done. In Ancient Egypt, other deities related to the "sea monster" (Leviathan) are the Hippopotamus deity of "**child birth**" (**brainchild**, created ideas or manifested imagination) called

Tawaret, the Crocodile deity of the Nile called **Sobek**, and the serpent called **Apep**. The Hippopotamus Tawaret and the Crocodile Sobek are combined into the Egyptian deity called **Ammit** who symbolically **"devours souls"** if a person fails the test of **Ma'at** and **Tehuti** (truthful or logical thought) in the **Underworld**. Another name for Leviathan is **"the Sea Serpent"** or "the **Dragon**" which is called **"Tanniyn"** in Hebrew, and the word **Tanniyn** is phonetically similar to **Tatenen**, the African deity that rose as the creator and **primordial mound** from the **chaotic waters** (Hell) called **Nun**. As ancient Alchemy gave way to modern Chemistry, this Nun, "Chaotic Water," "**Fire-Water**" or flammable liquid became known as **Naphtha** or **Napalm** (which stands for **NA**phthenic acid and **PA**lmitic acid). In modern chemistry, Naphtha would include substances like gasoline, oil, alcohol, and **Ether**. The word "**Naphtha**" was associated with the deity in Hinduism and Zoroastrianism named "**Apam Napat**" and the Aramaic word **Naphta** coming from **Napata** or **Na-Ptah** meaning **"the people of Ptah**," dwellers of Memphis, the blacksmiths and craftsmen. This Naphta is also associated with the Biblical character named **Naphtuhim**, son of **Mizraim** (which means the **two Egypts, Upper and Lower**) as well as the word used for a **Blacksmith** or *"one who transforms the elements"* (applies knowledge) in ancient Mesopotamia, called **Nappahu**. Also, within Ancient Mesopotamian culture, "The Science of Sciences" principles are depicted in eight-pointed, octo-grams (squared-circles or sun-crosses) representing the **celestial "Sky"** (symbolic of **"the Mind"**), and **"the Sun"** (symbolic of

Empiricism), can be found in the glyphs of the names for the Sumerian deity **Anu** (the sky) and the Akkadian deity **Shamash** (the sun).

Sumerian octo-gram cuneiform for the sky deity Anu	Akkadian octo-gram for the sun deity Shamash

In Mesopotamia culture, the ability to "**transform**" or "**cross the chaotic waters**" from the Rational to the Empirical worlds was called **Nibiru** in mythology.   Nibiru means a "**point of transition to cross the waters.**"   Modern speculations about Nibiru call it "**Planet X**" (with the letter X being a "**cross**"). Nibiru was considered the seat of the Babylonian deity Marduk who was a deity associated with both water and the sun being called AMAR.UTU meaning "bull of the Sun, Shamash."   The son of Marduk was called **Nabu** who was a deity of **wisdom**, which was obviously a mythological portrayal of the fact that "wisdom", or the ability to apply knowledge, is a result "crossing" the waters, or making the transformation from the Rational world to the Empirical world.   The Egyptians represented the changes or **transformations** that take place in the Empirical world with the deity **Khep-Re** (also spelled **Khepri**), and the changes that occur in the Rational world with the deity **Khonsu**, "**the traveler**".

Khepri	Khonsu
Empirical Transformation	**Rational Transformation**

The Empirical Transformation or "Movement of the Sun" in the "Science of Sciences" of Ancient African Culture and Philosophy had 4 major points: The **Rising Sun**, The Sun at **High Noon**, The **Setting Sun**, and The Sun in the **Underworld** (**Night**). These 4 positions of the sun or "types of scientific empirical experiences" have had various names throughout Africa and at different points in time in Ancient Egypt. The Rising Sun, High Noon Sun, Setting Sun, and Unseen or "Underworld" sun have been called "**Khep-Re, RE, Atum, Amun**" by some, and "**Tum, Tun, Mun, Nun**" by others respectively as the sun would travel in a circular motion from east to west across the sky. Each "position of the Sun" was represented by a different "deity" or principle. The deity **Ptah** was also synonymous with "**The North Pole**" on a **directional compass**, or "**The Sun at high noon**," and the "**pygmy sons of Ptah**" represented different degrees of the daytime arc of the sun across the sky. Another image of the 4 movements of the sun in Ancient Egypt can be found during the Ptolemaic period.

4-moments of the sun	The Sun between the two Horizons

In art, the point where the Sun would be "born" in the east from the underworld, and "die" in the west and return to the underworld on it's routine cycle was depicted by two Lions protecting the Horizon called "**Yesterday** and **Tomorrow**" who were the "guardians of knowledge"; another reference to the Empirical and Rational duality of "Light" or information. One of the earliest depictions of this concept comes from the 21st Dynasty **Papyrus of Dama Heroub**, where the two lions of the Horizon are shown facing east and west with the sun Horus on a lotus surrounded by **Ouroboros** in the center. Thus, the depiction of the Ouroboros predates Asian or European use, and the actual description of the Ouroboros is even older from the 5th Dynasty pyramid of **Unus** where another form of the duality is written in the inscription: *"A serpent is entwined by a serpent, the male bites the female, and the female bits the male, heaven and earth is encompassed."*

Considering that these four movements of the sun are different types or "degrees" of experience, then the true "serpent" is in fact the Mind which includes the Central Nervous System of the Brain and Spinal column. When the sun or empirical experience died or ended (the event occurred and was over) the sun traveled into "the underworld" or the Rational world (the world of the mind to be remembered and reflected upon) which was symbolized by the moon.

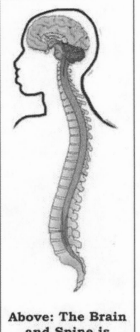

Above: The Brain and Spine is "The Serpent"

Just as there were **four moments of the Sun** or **Empiricism**, there were **four moments the Moon** or **Rationalism**. The four positions of the moon were the **waning crescent**, the **waxing crescent**, the **waxing Gibbous**, and the **waning Gibbous**. Egyptian deities that related to the moon or Rationalism were **Tehuti, Khonsu, Osiris**, and **Aah**. Even the deity **Nun** who represented the **"primordial waters"** was somewhat associated with the Moon because the gravitation of the Moon does pull on the tides of **the Oceans**, but also so does the gravity of the Sun.

The unification of the four positions of the sun, and four positions of the moon, which is symbolic for Empiricism and Rationalism respectively, is the basis of all science. "The

Science of Sciences" is Empiricism and "The Science In Sciences" is Rationalism. The Full moon on the bottom, or South, of the diagram would represent the complete "death" of the sun, whereas the sun at the top, or point "north," of the diagram would represent the complete "death" of the moon.

Above: Hapy performing sema-tawy Unification

The actual unification of these two dualities or "Lands" called Empiricism and Rationalism in "The Science of Sciences and The Science of Sciences" was called **Sema Tawi** (or **S.M.T.** in medu neter) in Egypt and was an **act** said to have been performed by the Egyptian deity named **Hapi**. This "Sema Tawi" Act or **Method** of unifying the Empirical and Rational world is also the name for **Egyptian Yoga**. **Pa-Seshat-Tawi** is "the Division of the Two Lands."

Above: The Ankh

In Hinduism, Yoga is a Physical (Empirical) practice that is said to help mediate the mind (Rational), hence, it is intended to "unite the two lands." The word "Yoga" comes from a Sanskrit root "**yuj**" meaning "to yoke" or "**to unite**." Thus, it is clear to see the African origin of the aforementioned Asian philosophies. Symbolically, the unification of dualities was depicted in the symbol known as **the Ankh**.

The Ankh symbol is said to represent "life" as it is the unification of the "male" and "female" principles of duality. Philosophically, as it relates to the **S.O.S.A.S.I.S.**, the unification of the dual principles of masculine and feminine that are depicted in the Ankh symbol would also represent **"the circle and the square**," "the infinite and the finite," and also **"the empirical and the rational**" respectively. As it relates to humans, the unification of these dual principles is called **"Sex"** and also **"Nuptials,"** and is the key to **creating** life. Also, the unification of these dual principles of the rational mind probing into the empirical world is called **"Science"** and does indeed lead to continuous creation. Life is sustained by continuous problem solving (Liberation) and death is the accumulation of problems unsolved. It is no surprise then that the etymology of the words **"Sex"** and **"Science"** both mean "to divide, **separate**, or cut" which of course relates to the African Creation Deity **PTAH** "the opener, or separator" (the deification of the Science of Sciences). Considering that the S.O.S.A.S.I.S. philosophy and theology of Egypt had to originate somewhere, and that the deity Ptah was acknowledged in Pre-Dynastic Egypt, then it is reasonable that the "Science of Sciences" may pre-date Egypt and could have quite possibly originated in the location where Egyptians said they originated from, at the source of Nile river amongst the pygmy tribes in Central Africa. Also, the origin of the Ankh symbol can be traced to an ancient cosmogram symbol used by the pygmy tribes at the source of the Nile in the Congo.

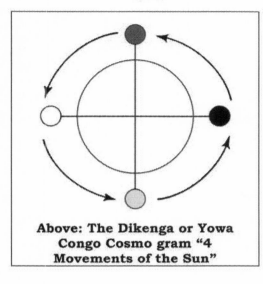

Above: The Dikenga or Yowa Congo Cosmo gram "4 Movements of the Sun"

The Congo cosmogram called the **Dikenga** or **Yowa** is said to be the origin of the Egyptian Ankh symbol (the unification of dualities and "key to eternal life"). The Congo cosmogram depicts the **"Tendwa Nza Congo"** meaning **"The Four Movements of the Sun."** This symbol is said to be the most ancient symbol as it was the first geometric shape given to Human beings by observing nature and the motion of the Sun as it traveled across the sky. The transformation or movement through each point of the Dikenga cosmogram is called "dingo-y-dingo" which means *"coming and going from the center."* Going from East to West, or counterclockwise like the motion of the sun, the 8 Phases of the Tendwa Nza Congo are:

1. **Kala**: morning, birth and childhood - colored black
2. **Kala Rising** (symbolic Summer)
3. **Tukula**: noon, the prime of life - colored red
4. **Tukula Falling** (symbolic Fall or Autumn)
5. **Luvemba**: sunset, late life, old age, death - colored white
6. **Luvemba Falling** (symbolic Winter)
7. **Masoni**: midnight, rebirth, resurrection - colored yellow
8. **Masoni Rising** (symbolic Spring)

The simple act of observing the sun and the moon traveling across the sky was the early empirical scientific observation in Africa that led to the development of the Dikenga symbol which not only represented the movement of the sun, but also the movement of "light" or information or knowledge, and thus is the earliest symbol of S.O.S.A.S.I.S method. The empirical observation of the sun traveling across the sky from east to west was evident. However, the human mind had to use the rational world of reason to construct the bottom half of the Dikenga "sun cycle" because the movement of the sun was not observed at night, but it was reasonable to assume that the cycle of the sun continued at night because the sun did "resurrect" and rise every morning. The circle of the sun was the first geometric shape; however, it was by observing the various phases of the Moon at night that led to the realization that the circle could be **divided**. This observation of the phases of the Moon gave birth to **the invention of the square**.

This is why the circle and the sun are associated with the empirical world (observation and experiences, etc.), and the square and the moon are associated with the rational world of thoughts and reason. Thus, the word "**crescent**" and the word "**create**" both share etymological origins because the crescent or moon,

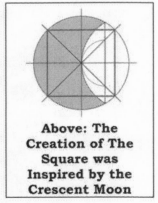

Above: The Creation of The Square was Inspired by the Crescent Moon

which is symbolic of the rational world, is the creative world that gives birth to "the sun" symbolic of the empirical world.

The S.O.S.A.S.I.S method and symbolism in Africa gave birth to a variety of similar philosophies and methods throughout the world. One of the most obvious philosophies that was derived from the African Dikenga is the speculative or philosophical side of **Alchemy** and **Hermeticism** which is supposed to lead to **"Ultimate Wisdom"** (Science, Useful Knowledge) by performing the **"Magnum Opus"** (or **Great Work** or **Great Method**) which leads to **"immortality"** (eternal life, continuous creation, liberation, and problem solving). It is apparent that not only the terminology, but also the colors and phases of the Alchemical Magnum Opus are derived from the African Dikenga. The four Stages of the Magnum Opus are:

- Sol Nigredo (Black Sun): change, dissolution
- Sol Albedo (White Sun): burnout of impurity
- Sol Citrinitas (Yellow Sun) : enlightenment
- Sol Rubedo (Red Sun): unity of the finite with the infinite

Even the **Ouroboros** (the serpent Wadjet) symbol which has it's origins as a symbol for the S.O.S.A.S.I.S. method in Egypt became adopted in Hermetic and Alchemical iconography as the **Caduceus** which is used as a symbol for modern medicine, doctors, and individuals who heal (Liberate, Problem solve, troubleshoot) ailments with the S.O.S.A.S.I.S. method; and

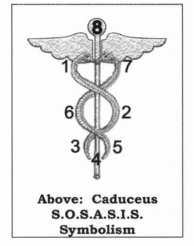

Above: Caduceus S.O.S.A.S.I.S. Symbolism

the Caduceus symbol depicts 8 Points as well.

Above: Narmer Palette

The first image of this Caduceus symbol comes from the Pre-Dynastic period in Egypt where the cobra Wadjet is depicted wrapped around a papyrus stem. Similar iconography of entwined "dragons" is depicted in the **Narmer palette** of **Menes** when he was the first Pharaoh to **"unite Lower and Upper Egypt"**. The Caduceus symbol as a staff is depicted in Egypt as two staff coiled by a serpent wearing the crown of upper Egypt and a serpent wearing the crown of lower Egypt being held by the deity Tehuti.

The doctrine of Tehuti is said to be one of the origins of modern Hermetic and Alchemical concepts. The letters in the abbreviation

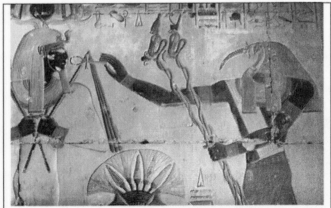

Above: African deity Tehuti (Thought) holding two coiled serpents, (the two lands) Empiricism and Rationalism

S.O.S. and S.I.S. geometrically form the coiled serpents of the Caduceus and also the figure-eight (**8**) symbol, and also the symbol for infinity (∞).

The African S.O.S.A.S.I.S. method also gave rise to many other Operative and Speculative symbols and tools in Africa and around the world. One of the obvious operative tools derived from the S.O.S.A.S.I.S. symbols is the navigational Compass with the four cardinal directions North, East, West, and South (abbreviated **N, E, W, S** respectively), and the four intermediate directions NE, NW, SE, and SW. Much like the S.O.S.A.S.I.S. with its 8-points helps the individual gain **direction** and **Navigate** the **flow of information** or "**News**" to the mind, the compass with its 8-points assists the user in gaining physical direction and Navigation in the air, on land, or sailing the **chaotic waters**. In an arc going from east to west, Ptah is considered the "North Pole" and the pygmy son's of Ptah are the various degrees in the arc. This principle is also found in the "Triangle Level and Plumb-Bob" tool used for balance and leveling when operatively building, and has often been used as a speculative symbol of "balance" or "meditation" in the S.O.S.A.S.I.S. philosophy.

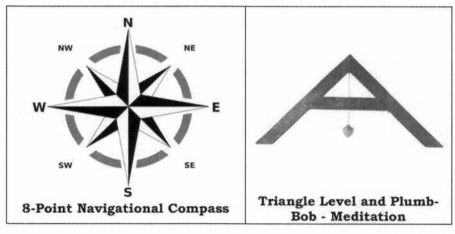

| 8-Point Navigational Compass | Triangle Level and Plumb-Bob - Meditation |

In other philosophical and religious uses, the Congo Dikenga cosmogram of the S.O.S.A.S.I.S. has been used in various **Voodoo** religious **beliefs**, and **Hoodoo** "magic" **practices**. In the West African **Ifá** wisdom divination system which seeks to "answer problems", "The Four Movements of the sun" are depicted on the opon Divination Tray of the Babalawo performing the ceremony. The word **"Ifá"** means **"knowledge"** (**Science**) and is also spelled and pronounced Ife, Nufa, Nufe, **Nupe**, or Nufawa in the Afro-Asiatic languages spoken by the **Hausa** tribes in West Africa.

The African **Ifá** system is similar to, but pre-dates, the Asian Zhouyi and Ba-Gua "8-fold path" system. The principles of the West African **"Ifá"** or "knowledge" (Science) system were personified in the various deities of the Orisha pantheon and their associated symbolic meanings and stories such as **Olokun**, **Obatala**, **Orunmila**, **Eshu**, and **Olodumare**.

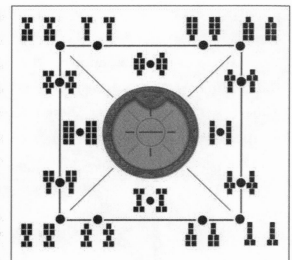

Above: Opon Tray used in West African Ifá (knowledge, Science) systems which depicts the 4 movements of the Sun (S.O.S.A.S.I.S.)

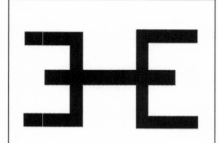

Above: The Kanaga S.O.S.A.S.I.S. Symbol used by the Dogon of Mali

The S.O.S.A.S.I.S. also appears in the philosophy and culture of the **Dogon** tribe in Mali. The sacred cosmogram of the Dogon is called the **Kanaga**, which shares similar symbolic meanings as the **Dikenga** symbol. The Dogon call "reality" or god by the name **Amma**, and it is part of the symbolic story that Amma created the **Nommos**. In the Dogon culture, the Nommos are **four sets of twins** (symbolic of duality) which total **8 beings** who are called "**teachers**".

The name "Amma" is called **Nyame** by the **Akan** people of West Africa, and the **duality of Nyame** is expressed in the **"Gye Nyame" Adinkra symbol**. The Akan people of West Africa express the importance of wisdom, and knowledge (Science) and the process of learning (The Science of Sciences and the Science in Sciences) in various stories, parables, and Adinkra symbols used throughout their cultural philosophy.

Above: Gye Nyame West African Adinkra Symbol Unifying the Dualities of existence

	Abode Santann	"The all seeing eye of the Divine Creator" Symbol of the totality of the universe, union of natural and social creations.
	Ananse Ntontan	"The Spider's Web" Symbol of wisdom, craftiness, creativity, and the complexities of life.
	Dame Dame	"Board Game" Symbol of strategy, intelligence, craftiness, and Reasoning
	Hwehwemudua	"Measuring Rod" Symbol of excellence, perfection, refinement, and superior quality.
	Mate Masie	"I have heard what you have said" Symbol of understanding, wisdom, knowledge, prudence, thoughtfulness.
	Nea Onnim	"He who does not know can know from learning" symbol of knowledge, and the pursuit of life-long learning
	Nyansapo	"Wisdom Knot" Only the wise know how to unravel it. Symbol of wisdom, ingenuity, intelligence, and patience.
	Osiadan Nyame	"God the Builder" Symbol of God the Builder and Creator of the World.

Above: West African Adinkra Symbols Related to "The Science of Sciences and The Science In Sciences"

These various practices, systems, and methodologies (Yoga in the Hindu culture) of obtaining "knowledge, understanding, wisdom, and enlightenment" to be in harmony and balance (meditation) with Nature and Reality (God) and solve problems (Liberation) have taken on various Operative and Speculative (Empirical and Rational) forms in a variety of traditions.

Various S.O.S.A.S.I.S. "Meditation" practices for problem solving

| Operative-Speculation | Speculative-Operation | The Speculation to Operation cycle |

The physical "postures" associated with "Meditation" actually do enable clarity of the mind in order to solve a problem, achieve a goal, or realize a benefit. However, the physical practice without "reason" is merely a "ritual", and the mental practice without physical action is idleness. The "Science of Sciences and the Science in Sciences" is what separates and unites **"Speculative Operation"** and **"Operative Speculation"**. Again, the unification of the Empirical "Science of Sciences" with the Rational "Science in Sciences" enables the ability to turn ideas into reality, or bring the thoughts in the mind symbolized as the "top" or "celestial", down to the real world symbolized as the "terrestrial". This creative ability to bring concepts and ideas from the mind into reality in Hinduism was called being an **Avatar** which symbolically meant "**descent**" or "**from heaven to earth**" (to bring from the top, down). This creative ability was symbolically depicted in the African S.O.S.A.S.I.S. stories associated with **Ptah** by saying that "*Ptah as Tatenen walked with his head in the skies and his feet on the ground*".

Studying the "Science of Sciences and the Science in Sciences" dictates that the process of "speculation" precedes the process "operation". Thus, the speculative nature of one's "way of life", cultural practices, religion, theology, cosmology, and philosophy, shapes and forms the mind and mentality as the base of all assumptions, hypothesis, and beliefs about reality and living. Hence, a "scientific-based" culture, philosophy, or "spiritual science" should lead to great success when moving into the operative mode of using and applying one's beliefs. Therefore, it is no surprise that Asian and Indian cultures that have a "scientific-based" philosophy and "way of life" also excel in achievement in the operative fields of Science, Mathematics, and Technology. In the USA, scholastic achievement in Science and Mathematics is correlated with money, socioeconomic status, and resources. However, even with all of the financial resources that the USA has at its disposal, the USA is still out performed in Science and Mathematics by Asian and Indian countries and other Industrialized Nations with similar resources. If scholastic achievement in Science and Mathematics is ever analyzed by the pupil's "cultural philosophy" or "spiritual science", it would not be surprising that the achievement of pupils with "scientific-based" philosophies and religions would far exceed the achievement of pupils with non-scientific-base philosophies and religions. Because Science and Mathematics are the operative expressions of "Knowledge" and "Understanding" that lead to the creation and development of "Technology" for survival and well being, then having the "Science of Sciences and the Science In Sciences" as a philosophical base for African culture should definitely lead to the creation, development, and production of African Technologies that will greatly improve the quality and standard of life for African people worldwide.

At this point, it is fit to discuss the term "Africa". When the term "African people" is used in this presentation, what is meant is the group or "Race" of people that are currently called the "Black Race" which has it's origin on the landmass that is currently called Africa. Even though both the "Race" of people and land mass called currently called "Africa" has had many other names throughout time, the term Africa currently and immediately identifies both a land mass and race of people in the minds of most readers. So as not to confuse readers or go off on a tangent, the term "Africa" is used for clarity purposes based on the reasons stated.

Moreover, etymological studies of the word **"Africa"** usually result in either Africa coming from the Afro-Asiatic word **"Afar"** meaning dust, or from the Arabic word **"Faraqa"** meaning "to divide or separate". In the case of the word "Africa" coming from the word "Afar," considering the morphology of Afro-Asiatic words where the "F" sound is similar to the "P" and "Ph" sound, then it is reasonable to see how the word "Afar" meaning "Dust" in an Afro-Asiatic language can be derived from the word "**Ptah**" which refers to the "Earth (or dust)" in traditional African language. Also, in mythology, "Ptah" is considered "**the Opener**" which implies "**separation**" (Science). In the case of the word "Africa" coming from the word "**Faraqa**" meaning "**to divide or separate**", this word would be synonymous to the etymology of the word "**Science**" which means "**to separate**". Even if one were to argue that the word "Africa" was derived from the name "**Scipio Africanus**," the etymology would still lead to the name "Scipio" (meaning '**to cut**" or **separate**) and "**Africanus**" (derived from the Afro-Asiatic word "Afar") having the same etymological meaning as the word "**Science**." In any case, the word Africa is used within this presentation because of its positive associations to a particular "Race" of people, as well as to Science and Ptah (The Science of Sciences and the Science in Sciences deity).

Also, the term "Race" is used reluctantly because "Racial" classifications are extremely subjective, speculative, and non-scientific qualifications. Most "Race-based" beliefs and hypothesis fail when tested scientifically using Biology and Genetics. However, for the purpose of understanding, the term "Black Race" or "African Race" is used because those terms readily identify the group of people that is being referred to in the mind of most readers.

This book entitled "The Science of Sciences and The Science in Sciences" is book 1 of a project which has been entitled "The African Liberation Science, Math, and Technology (abbreviated S.M.A.T.) Project." The purpose and intention of this book "The Science of Sciences and The Science in Sciences" is to: 1) describe the method by which information is acquired, processed and applied; 2) to depict in words, forms, formulas, and symbols The Science of Sciences and the Science in Sciences; and 3) to show the origin of the Modern Scientific and Mathematic method in Ancient and Traditional African culture and philosophy. This book entitled "The Science of Sciences and The Science in Sciences" intends to motivate the African reader to embrace their traditional "scientific-based" culture and philosophy so that African people will become **creators** of highly efficient and advanced operative Sciences and Technologies for survival, well being, problem solving, liberation, healing, and positive standards of living and quality of life. This book is the result of years of experience, observation, introspection, and contemplation about existence in general and the meaning and purpose of life in particular.

A problem is an obstacle which makes it difficult to achieve a desired goal, objective, purpose, or "any thing that opposes, resists, or challenges the act of going from a current state to a desired new state". The act of going from a current state to a new state is the definition of "**Change**." "Change", for better or for worse, is what causes "**Creation**". The word "**Liberation**" means to be able to **act** as **desired** without obstruction, and thus, "Liberation" is synonymous to "Problem Solving" and required for Creativity. Creativity is accomplished by "uniting" the "two lands" of the "mind" and "reality". In Egypt, the unification of "the two lands" was called **Smai Tawi** or **S.M.T.** and was performed by the deity named **Hapy** in Egyptian mythology who was symbolic of the flooding of the Nile River which enabled the ability to **"grow crops"** (create) called **Nu-Pu-Re** or **Nepri** (Grain or Corn) in Ancient Egypt. Much like the ancient African practice of **"unifying the two Lands"** called **Smai Tawi** or **S.M.T.**, African Creation Energy's African Liberation **Science**, **Math**, and **Technology** (abbreviated **S.M.A.T.** or **S.M.T.**) is the unification of Science, Math, and Technology. Of course, "Unification" occurs after separation or "Division of the two lands" which was called **Seshat Tawi** in ancient Egypt, and **Seshat** was a goddess of Science, Math, and Technology who surveyed the land for building after the flooding of the Nile. The principles of **Science**, **Math**, and **Technology** are the operative expressions of the philosophical terms "**Knowledge**, **Understanding**, and **Wisdom**" respectively. Thus, the Africa Liberation S.M.A.T. project enables and empowers those who read, comprehend, and apply the information with the creative ability to **"grow mental crops"** and turn the concepts and ideas in the mind into reality. The African Liberation S.M.A.T. project is designed to actively provide useful information to help **"Liberate"** or **"solve the problems"** of African people. The African Liberation S.M.A.T. project unifies the "two lands" of the Rational Mind and the

Empirical world enabling proper comprehension, explanation, and reasoning of empirical experiences from the real world into the rational world, and proper transformation of rational thoughts, concepts, imaginations, and ideas from the world of the mind into the empirical world. This is the cycle of creation, every thought is experienced, and every experience is reflected upon or thought about. After an event has happened, that event is symbolically "dead." However, any observers to the event can "reflect" upon or think about their relative experiences with the event much like the moon reflects the sun's light. And, if someone chooses, they can re-create or give "birth" to a similar event as the initial one they experienced, or the initial experience can give birth to ideas in the mind that result in creating new experiences similar, but unlike the initial experience. Information is the source of Knowledge, because you have to obtain information in order to know you are in a current state, and you have to have information to know you want to go to another state, and the union of these two pieces of information enables you to know you have a problem. Moreover, you need information in order to "Liberate" or solve the problem. A lack of information to solve-problems (liberation) leads to the problem persisting, and thus suffering. Creation, or the change of states between the dualities of existence, is perpetual. Thus, Liberation or "problem solving" is requisite for creation and Life. Since **liberation is "problem solving**," then a real Liberator is one who teaches people to solve their own problems. As long as someone is "Liberating" or solving your problems for you, you will never be free, because you will be dependant on your liberator or "problem solver" or **savior** for solutions. As Argentinian revolutionary **Che Guevara** said, "*I am not a Liberator, the people liberate themselves*". A true Liberator is someone providing **"Liberation Information"** which instructs or teaches problem solving. It may be necessary to provide results for a short interim period of time while "problem solving" is being taught, but once the pupil has demonstrated the ability to solve problems or "Liberate themselves" (wisdom), then the providing of results is no longer necessary. This concept was symbolically depicted as **"being resurrected from the sarcophagus of death back to life"** in ancient African philosophy, mythology, and iconography. Within this book entitled **"The Science of Sciences and the Science in Sciences"** the reader will experience information which provides **"Scientific Liberation"**, that is, information that provides a means to answer any and all questions, solve any and all problems, and respond to any and all calls that are ever encountered. This "Preface of Science" has discussed the African cultural and philosophical origin of the methods that are known as **"The Scientific Method"** and **"The Mathematic Method"** which are collectively referred to in this book as **"The Science of Sciences and The Science of Sciences"**. Indeed, "The Science of Sciences and The Science of Sciences" is the foundation of Science, the Origin of Knowledge, the Key to Problem Solving, Liberation, the "Way of Life", and the way by which Life and all of creation is created.

2.0. THE INTRODUCTION OF SCIENCE

Science is defined as a system, method, technique, or process used for acquiring knowledge and information based on experiences with natural phenomenon. Scientific knowledge can be used to explain and predict the causes, effects, and composition of natural phenomenon. Science has many fields, classifications, disciplines, and categories. Modern Scientists classify scientific fields into 3 categories: Natural Science, Social Science, and Formal Science. The Natural Science category includes the fields called Physics, Astronomy, Biology, Earth Sciences and Chemistry which all deal with studying Nature. The Social Science category deals with scientific fields that study Human behavior and society which includes the fields called Anthropology, Archaeology, Philology, Communication, Economics, Education, Law, Linguistics, Politics, Psychology, and Sociology. The Formal Science category studies Logic, Mathematics, Computer Science, Information Theory, Decision Theory, Statistics and other Formal Systems based on definitions and rules. One major error in the Modern Classification of Scientific fields is to separate the Natural Sciences from the Social Sciences as if to suggest that Humans and Human Behavior and Society is separate from Nature. Science is the unification of theoretical knowledge (called **episteme**) and methods for practical application of knowledge to produce results (called **tekne**).

The word "Science" is perhaps one of the most powerful words in the English language. The word "Science" was not invented until about 1300 A.D., but the etymological origin of the word Science is much older. Originally, the word "Science" meant "knowledge acquired by learning or studying," but it was derived from the Latin word "scientia" and "scire" meaning "knowledge" and "to know", which was derived from the older proto-indo-European word "skei" meaning "to separate, to distinguish, to split, cleave, rend, or divide." The concept of "dividing and separating" when it comes to knowledge was depicted in the Ancient African stories of the deity **Ptah**, who was called **"The Opener"** or "the Creator" who rose or **"separated"** from the chaotic waters (fire-water) of **Nun** as "risen land" (or pyramid) called **Tatenen**. In West Africa, these concepts are found in the name of the **Dogon** tribe of **Mali** where the word Dogon means **"People of the High Mountains,"** and they say that their ancestors who are called **Nommo** meaning **"Master Teachers of the Water"** came from a **Binary Star** constellation named **Sirius** (called **Sigui Tolo** and **Po Tolo** to the Dogon). This concept is found in the Jewish religion as a place in Jerusalem used for the name of the city of **David** (Daoud in Aramaic) called **Mount Zion** or **Scion**, from the Hebrew **Tsiyown** (Strong's H6726) which meant "a **High Sunny and Dry (free of water) Mountain.**" The word **"Scion"** also refers to **"the new growth of a plant"** and has etymological meanings of **"to sprout, split**, and **open**," which obviously is related to the African deity Ptah, "the opener" and deification of **"The Science of Sciences and the Science in Sciences"** (abbreviated S.O.S.A.S.I.S.). The abbreviation **S.O.S.A.S.I.S.** does also appear to look like the word **"S.S. OASIS"** where "S.S." is an abbreviation for **"Steam Ship"** in modern nautical terms, and "Oasis" is considered a type of "heaven," and we do indeed see the S.O.S.A.S.I.S. method as a **Path** or "ship" to get to symbolic **Heaven** (Liberation) to leave from out of the "Lake of Fire" called **Hell**.

In the abbreviation S.O.S.A.S.I.S., the **S.O.S.** portion of the abbreviation does stand for the **"Science of Sciences"**, which is the Scientific method of processing or **Navigating Empirical Information** or **News**. In modern usage, the letters **SOS** are used as a **Morse code signal** consisting of 9 marks, *"3 dots, 3 dashes, and 3 dots,"* which indicate **distress** and the **need for help** and **assistance**. It is said that when used in **nautical** terms, SOS stands for "Save Our Ship," "Save Our Souls," and "Sink Or Swim," and indeed, the "Science of Sciences" or S.O.S. method does enable the user to "save their soul (mind) and ship (body)" as the compass (North, East, West, and South or N.E.W.S.) to successfully navigate the flow of the metaphorical **sea** "waters of information and news." The "Science of Sciences" or S.O.S. does provide assistance, help and direction as the mental **Navigator** and **Path-finder** for the best course of action. In the abbreviation S.O.S.A.S.I.S., the **S.I.S.** portion of the abbreviation does stand for the **"Science In Sciences,"** which is the Scientific or Mathematic method of **processing** and **Reasoning Rational Information**. In modern usage in the English language, the letters **SIS** are used as a shorter form of the word **"Sister."** In the scientific field of Biology, the word "sister" is used to denote a **companion** or second cell formed by division, and indeed the "Science In Sciences" or S.I.S. is the **"sister"** or companion to the "Science of Sciences" or S.O.S. Also, the word "Sister" is related to the ancient concept of the "SI-STAR," where the word **"Si"** was an ancient Mesopotamian deity of The Moon also called **Sin** or **Su'en** or **Zu.En** "the lord of wisdom." This deity Su'en or Zu.En is also related to the word **Zion** as previously discussed and is associated with **"the Serpent"** in Monotheistic religions. The deity of the Moon and Wisdom (Rationalism or the Mind) called SI or Sin was also associated to the **Heart** and **Mind** of **Ptah** called **Sia** or **Saa**. The Egyptian word for heart is **"Ab,"** thus Sia or Saa as the "heart of Ptah" was called "Ab Saa" which was associated with the Mesopotamian concept called **Ab.Su** "the chaotic waters (lake of fire)" called the **Abyss."** The Rational mind (symbolized by "the Moon" in ancient times), that reflects on the "Light" (Empirical information symbolized by the sun in ancient times) can be both positive and negative. The mental process of reason and Imagination can be used for the purposes of Creating means of survival, well being, solutions, Liberation, and "Heaven;" or the Imagination process can delude, and the "SI-Star" can create "Moon Stars" or "Monsters" and "Devils" (serpents) that can plague and torment the mind of an individual leaving them in an endless abyss of speculation and "Hell." In "The Science of Sciences and the Science In Sciences," the "sister" called Rationalism or Reason (the mind or moon) is united with Empiricism and Experimentation (the Sun). All thoughts and ideas generated by the "sister" called "SIS" or "the Science In Sciences" are checked by means of Logic, Reason, (Mathematics) Evidence, Experience, and Experimentation (Science) to ensure that the mind gives birth to Ideas, thoughts, Imaginations, and Creations (technologies) that are in tune and aligned with Nature. Thus, in the S.O.S.A.S.I.S., "The Science In Sciences" is based in "Reason" and tested in "Reality" and used in the positive sense.

In order to provide the many attributes and meanings for the word SCIENCE, we use a "Letter-by-Letter" acronym for the word "Science" that, when combined with the definition and etymology of the word "Science," gives a broader scope of the multi-faceted concept called "Science." The S in the word "Science" stands for Separate, Split, Slice, Ski, Scribe, Shed, Shit (fertilizer), Sex, Skeptic, Scorpion, Scepter, Staff, Scimitar, Sword, Shield, Shell, Secrete, Secret, Sacred, Section, Solutions, Sense, Sail, as all of these words have meanings and origins related to separation and division which is requisite for Science and allude to the creative abilities of Synthesis needed for Survival; and both Synthesis and Survival are used to represent the S in the word Science. The S also stands for **Sia** or **Saa**, the heart and mind of **Ptah**. The S stands for "Something-ness" or matter, and "Space" or vacuum, which are both studied in the Scope of Science. The S in "Science" also stands for "Solving Secrets (mysteries)" and "Spiritual Séances" because Science does indeed deal in Specifics and clarifies Speculations which in turn helps establish Sanity and makes the individual Secure and not Scared and Silent. The S in "Science" represents the Scenario where one becomes a "wise Serpent" (or Snake) and Sapient Sage. The letter C in the word "Science" stands for Conscious and Conscience, as well as Consideration, Contemplation, Concentration, Criticism, Consequence, and Conclusion; as all of these C words are mental process that are required for Science. The letter I in the word "Science" stands for Information, Investigation, Identification, Innovation, Ingenuity, and Induction which are all causes or effects of the Infinite Intelligence obtained by Science. The next letter in the word "Science" is the letter E, which would be appropriate to represent Experiments, Experience, Elimination, Evidence, Exploration, Equilibrium, and Efficiency, which are all needed for Scientific Examination into Eternity, Entirety, Existence, Evolution, Energy, and Elements. The letter N in the word "Science" represents the scientific study of Nature and the Natural world. The letter N also stands for "Nothing-ness" (Null) which is Space or Vacuum, as well as Navigation, the mind called Nous (Noll), Nativity, and the Ancient African concept called Nun from which Science came. The second letter C in the word "Science" stands for the Characteristics of Confidence, Comprehension, and Creativity that result as a Continuous use of Science. Lastly, the second letter E in the word "Science" stands for Education, Engineering, and Enlightenment, which are all Exceptional results of the Excellent use of Science.

This section of the book entitled "The Introduction of Science" is intended to give the reader a clear description and definition of Science. So that this book is a book of Knowledge (Science) and not ignorance, we will answer all of the **"Knowledge Questions"** that pertain to the topic of Science. We have specified exactly **"What** is Science" and the "etymology" or origin of the word Science, but to gain further insight, we must examine the "Philosophy of Science." Indeed the "Science of Sciences and The Science In Sciences" which is called the "Scientific Method" and the "Mathematic Method" in modern terms, is considered the "philosophy of Science." However, certain definitions about knowledge, truth, existence, and Information must first be established before the methodology of Science is described in detail. Philosophy provides a reasonable and systematic means for addressing these topics, therefore, before we go into the specifics of the methodology (answering the **"How"** questions) that is the "Science of Sciences and the Science In Sciences," we will establish our definitions of the semantics and basic fundamental aspects and components required for the Philosophy of Science by examining the Nature of Science, the Ontology of Science, the Noology of Science, and the Epistemology of Science (answering the **Who**, **Where**, **Why**, and **When** questions). The topics of Nature, Ontology, Noology, and Epistemology will provide answers to the Knowledge questions about Science and will collectively form our "Philosophy of Science." Then we will describe the "Methodology of Science" that is called "The Science of Sciences and the Science In Sciences."

2.1. The Nature of Science

The word "Nature" means "essential qualities or characteristics." Therefore, to describe the "Nature of Science" we must present the essential qualities and characteristics of Science. In institutional Philosophy, the study of "The Nature" of beings is covered in the field called **Meta-Physics**. However, the word "Physics" means "Nature" and therefore the word "Meta-Physics" implies something that is "After Nature," "Beyond Nature," or "Super Natural." In the broadest sense, we define the components of Nature as Space/Vacuum, Matter/Energy, and Time/Existence. As a result of this definition of the component parts of Nature, there is nothing in existence that is "After Nature," "Beyond Nature," or "Supernatural" that would be outside and not apart of the component parts of Nature; to us, these terms are **oxymorons**.

In the Ancient African philosophy, we consider Nature as "The All" and thus nothing can be outside or apart from the all. There may be Observations, Events, Instances, and Phenomenon that are "Abnormal," rare, and infrequent; but not outside of the all. The term "paranormal" is used to refer to those observations in nature that do not occur frequently or regularly, or are not witnessed, observed, and experienced frequently or regularly; however, they are not above and beyond nature. These paranormal, abnormal, and infrequent phenomenon are still part of Nature as they can and would be classified into one or more of the component parts of Nature. So for the reasons stated, we do not discuss the "Meta-Physics" of Science, but present the Nature of Science as a means to describe the essential qualities and characteristics of Science.

As we discussed earlier, the word "Science" means "Knowledge" and the etymological origin of the word "Science" means "to separate." The information provided by the definition and etymology of the word "Science" gives insight into the "Nature of Science" which we will expound here. In describing the "Nature" or essential qualities of Science, we maintain that there are basically two parts to Science: **The Known and The Unknown**. This is one of the reasons why the etymology of the word "Science" means "to separate," because Science is composed of two separate parts. The parts of "Science" called "The Known" and "The Unknown" were symbolized in Ancient African Philosophy as "the Ptah-Nun" and "the Nun."

The Known and The Unknown aspects of Science can also be symbolized by the Finite and the Infinite, because things known are confined to the Finiteness of the Mind, and things unknown are Infinite. Geometric symbols for the Known-and-Unknown duality of Science are the Square and the Circle respectively. Gender symbols for the Known-and-Unknown duality of Science are the Man and the Woman respectively. Science is an attempt to make the unknown, known, and symbolically make the infinite, finite and "square the circle". Mathematics has shown through logic and reason that it is not possible to "square the circle" or make a square have the same area as a circle even though they both have 360 degrees due to the **"transcendental"** Nature of Pi (3.14...) within the Circle. The Infinite cannot be made finite, light cannot consume darkness, and all that is Unknown cannot ever totally be made known. However, it is the nature of the man to want to go into the woman. Science is symbolically the "Man" probing into the "Woman," (Sex) and since it is the Nature of the Man to want to probe into the Woman which leads to creation, it is also the Nature of Knowledge (Science) to want to know more. Symbolically, it is the Nature of "the Woman" to want to attract "the Man" which leads to creation, and it is also the Nature of Information to want to be experienced. Thus, it is the Nature of Science to lead to creation, development, and change.

The true and rather paradoxical statement that "The only Constant in Nature is Change" reflects the perpetual and endless Creation that takes place in Nature, because "Change" is Scientifically Creation. This need for constant "Change" or constant "Creation" is satisfied by the dualities in Nature which manifest into a multitude of forms. The separation of information into the dualities of known and unknown leads to the quest for knowledge, which leads to knowledge (Science) by way of comprehension, which in turns leads to wisdom or the use and application of the initial information to continue the perpetual cycle of creation and change in Nature.

In order for there to be Science or knowledge, there must be something to know. In order to know something, there must be an observer and something to be observed; again "the known" and "the unknown" duality which make up the essential qualities or Nature of Science. The "observer and the one being observed", or "the Known and the Unknown" can also be called "the one experiencing and the one providing the experience", "the one transmitting information and the one receiving information", or "the Teacher and the Student".

The reality is that the duality paradigm of Science is relative and while one side of the dichotomy maybe called "the Known" it is simultaneously "the Unknown" relative to an observer. This is best explained using the "Teacher-Student" analogy. In this Scenario, the "Student" or "pupil" represents "the seeker" of knowledge or "the observer" and the "Teacher" represents the one providing the knowledge or the one being observed. However, just as the Pupil is observing and learning from the Teacher, the Teacher is observing and learning from the Pupil. Relative to the Pupil, the Information or Knowledge (light or Science) must be obtained, absorbed or Comprehended before it can be used and applied (Wisdom). However, relative to the Teacher, the Teacher does not know that the student has comprehended the information until it is used. Thus the student looks for Knowledge and Comprehension as a means to Wisdom, whereas the Teacher imparts Knowledge and looks for Wisdom to know that Comprehension has occurred.

As we conclude this sub-section called the "Nature of Science," we have described that the "Nature" or "essential qualities" and components of Science are basically binary being "The Known" and "The Unknown." As the old saying goes, **"There are two sides to every story,"** and since EVERY story is contained in the Information within Nature, then there are also "two sides" to Science. It is the "weighing" or "judging" (determining) the information contained within the "two sides" of the Science story that determines Truth, and we will discuss the concept of truth in greater detail in the "Epistemology of Science" sub-section. This "weighing" or "judging" between the information contained within the two-sides of the Science dichotomy to determine "Truth" was symbolized in Ancient African culture and Philosophy by the **"Scales of Ma'at"** in the Underworld.

Since Science is based on Knowledge and Knowledge is based on Information, and Information is all of Nature (Space/Vacuum, Matter/Energy, and Time/Existence), then indeed the answer to the knowledge question about **"Where is Science"** is **"Everywhere."** The answer to the knowledge question about **"When is Science"** is **"Always,"** and the answer to the knowledge question about **"What is Science"** is **"Everything."** The purpose of Science is for **perpetual change and creation in Nature**, and this answers the knowledge question about **"Why is Science."**

2.2. The Ontology of Science

Following our discussion on the "Nature of Science" we will go forward with our discussion on the "Ontology of Science" in our "Philosophy of Science" presentation. Ontology is defined as the philosophical study of existence, reality, and the classification of "beings." And, by "beings" we mean anything that exists in the state of "to be," and thus are "beings;" this would include anything and everything in Nature (Space/Vacuum, Matter/Energy, and Time/Existence). Just as the "Nature of Science" is a dichotomy that we call "The Known and the Unknown," the Ontology of Science is dichotomous also. The "Ontology of Science" or the "Existence of Science" either exists as "Abstract or Concrete." We can also call this duality in the Ontology of Science by the labels, "the Rational and the Empirical," "the Subjective and the Objective," "the Speculative and the Operative," and the "Hypothetic and the Pragmatic" respectively. These dualities were also symbolized by "The Moon and the Sun" or "Death and Life" in Ancient African culture and philosophy. Science, knowledge, or Information either exists as a concept or as a "reality." We maintain that these are the only states that all information exists in, and thus these dualities are the only existences of interest when discussing the Ontology of Science. All of Nature or all of existence is made up of "Information," and it is Information that forms "Knowledge" or Science. Just as there is relativity in the "Nature of Science," there is relativity in the "Ontology of Science" in so far as what would be considered "Abstract" and what would be considered "Concrete" depends on the observer. Since Ontology is the study of "existence," then it would be appropriate to expound on the "Where," "When," "Beginning of," and "Ending of" Science in this section as these are knowledge questions that are answered by way of Ontology. As we stated early Science is Knowledge, and knowledge is Information, and Information is everywhere all the time. Information is Energy detected by Matter within Space and Time. In this discussion about the "Ontology of Science" we must delve into the concept of Time/Existence in Nature. Time is basically the use of one event to measure or quantify in multitude and magnitude another event. When most people think of time, they are referring to existence. For example, the whole time system as most people know it today is based on the event of the Earth rotating around its axis. The event is then divided down to the unit of time called the "second". In modern watches and clocks that are used to measure time, a quartz crystal is used due to a special property called the **piezoelectric effect** that enables the crystal to vibrate a certain number of times when a certain amount of electricity is applied to the crystal. This piezoelectric effect was approximately accurate enough to model the unit of time called the "second" and it is the piezoelectric effect that takes places in watches and clocks in **the space between the TIC and the TOC**. Therefore, in order for there to be time, there must be at least "two" things: an event to be "timed" and another event to time it by (again going back to the binary Nature and Ontology of Science). Also, time requires "events" and "events" are "change in state," thus, if nothing is moving and nothing is changing, then there are no events and there is no time; just a single long moment until change occurs. This is why we maintain that the concept called "Time" is a way to measure "Existence," because even in the case where nothing is moving and there are no events occurring, that single long moment would still be Existence. In modern Science, the condition where nothing is moving, and thus there is no time, is called "**Absolute Zero**." Although nothing is moving in the state of Absolute Zero, there still exists what modern Scientist call "**Zero Point Energy**." Also, since movement or change (creation) is considered "**Entropy**" or "**Chaos**" in modern Science, then the state of Absolute Zero where there is no time would also be considered the state of minimum Entropy, minimum Chaos, or **Maximum Order**. However, this is just a theoretical concept because Science has shown that the State of Absolute Zero cannot be reached. Furthermore, the 2nd Law of Thermodynamics states that the entropy (Chaos) in the Universe will continue to increase until equilibrium or **Balance** is obtained between all of the forces in the Universes of Nature. With this background scientific information, we can now discuss the Ontology of Science. Science as a concept began when information began, and information began when the Universe began. Science is everywhere anytime there is information. Modern Scientist say that the Universe began with "**A Big Bang**" (or **Big Separation**), and this theory is based on the observation that **the Universe**

is Expanding (the Gap in Separation is becoming larger and larger). This expansion is taking place at the Astronomical level, but will eventually lead to the expansion of the cellular, atomic, sub-atomic, and other "universes" as well. This expansion or constant Separation is Change, Creation, or **"Entropy"** in modern Science, and the reasonable conclusion of this expansion will be the complete separation of the **smallest particles of matter** (called **fundamental particles**) as far apart as they can get as they approach Absolute Zero; and this by some would be considered the "End of Science." However, since this state is theoretical and cannot be reached in practice, then eventually all of the particles of matter will begin to contract, combine, and **Unify** into what is called a **Singularity** in modern science; and this by some would be considered the "Beginning of Science" and the time before "time" or the moment before the "Big Bang." However, this state of singularity is theoretical also, but these theories dictate that the Universe expands and then contracts, and then separates and then unites in an endless cycle of creation and change. There is an attempt in modern Science to explain the separating and contracting duality of the Universe in the **"Dark Matter/Dark Energy" Hypothesis**. Modern scientists liken **"Dark Matter"** to **"that which unites the Universe"** and liken **"Dark Energy"** to **"that which separates the Universe"**. Even in the **"Dark Matter/Dark Energy"** Hypothesis, "Dark Energy" or "that which separates the Universe" is in greater abundance at this point in time.

Thus, there is no real beginning or ending of Science, but rather a cycle between the extreme dualities of Science, from Hypothesis to Experience

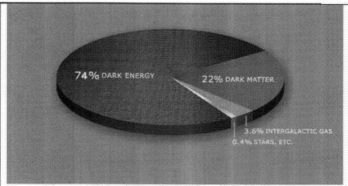

Above: The "Dark Matter/Dark Energy Hypothesis" suggests that currently the Universe is 74% Energy and 26% Matter. The Energy/Matter duality that this Hypothesis represents, cycles in quantity throughout existence with each one dominating a different points in time

over and over again. Considering this cycle of creation from experiences to thoughts and from thoughts to experiences, a reasonable question always comes up: Which came first the thought (hypothesis) or the experience? While this question is reasonable, it shows that the thinker is thinking linearly and not cyclically. All of creation is a cycle, and thus there is no first, only continuity. Thus, thoughts require experiences and experiences require thought; this is the transformation of energy into matter and matter into energy. Where modern scientist proclaim that **"The Big Bang,"** "Big Explosion," or "Big Separation" was the "beginning of the Universe", this is only half-true because creation is a cycle. If the universe is expanding, or in a stage of separation or **analysis** (which means **"to break apart"**), then prior to the big-bang (which was the beginning of the expansion of the universe) there must have been a **"Big Crunch"** which was the contraction, unification, combination, and synthesis (which means "to come together") of all the matter and energy in the universe into a singularity. After complete Unification, this singularity then "exploded" or began the process of separation and expansion called the "Big Bang". So indeed by **"The Science of Sciences and the Science in Sciences"** we can reason "what happened before the big bang." The expansion, separation, or analysis of the Universe is the transformation of matter into energy (experience into thoughts) and the contraction, unification, or synthesis of the Universe is the transformation of energy into matter (thoughts into experience); this is the cycle of creation. Complete separation or "Analysis" is the maximization and perfection of internal speculation; this is the condition that enables all of Nature to completely observe its smallest parts, and thus is a form of "introspection" or "Rationalism." This state of complete separation and "analysis" is also the imperfection or minimization of external speculation, external experience, external observation, or "Empiricism." After the complete separation period, the "Zero Point Energy" in the Universes of Nature begins to cause a contraction, synthesis, and pulling the Universes of Nature back together into a Singularity. The Complete Unification is the maximization of external observation (Empiricism) and the minimization of introspection (Rationalism). As all of Nature and all of Science alternates between these two extreme learning conditions, the many changes and creations that take place are considered a form of "chaos." One of the purposes for all of the creation and changes in Nature is so that all of Nature can learn about itself, and in order to do this, it vibrates between the two extremes of Hypothesis and Experience over and over creating many forms in between which serve the purpose of various experiments; this notion gave rise to the modern concept called **Determinism** or **Causality**. Both of these extreme conditions in Nature and Science are considered relative "order" at their respective times. The Universe, Science, and Nature approaches "absolute zero" in the state of complete separation as well as in the state of complete Unification or "singularity." Unification is the result of Learning, whereas Separation is the process of Learning. Everything that exists is an "Effect" of some "Cause," and this "cause" can be considered a "Hypothesis" in Science. However, every "Hypothesis" or "Cause" is preceded by some "Effect," and this "Effect" can be considered an "Experience" in Science. Thus, in the Ontology of Science, it is very important to identify if a piece of information exists as a "Cause or Effect," "Hypothesis or Experience," or "a Concept or a Reality." You cannot know something or be aware of something that does not exist, because by virtue of the fact that you "know it," then it at least exists as a "Concept" even though it may not exist in reality. On some level, however, something can exist that you do not know of, or are not aware of, and it is the pursuit of this awareness that is Science. Considering the Ontology of Science, and the fact that everything is always changing, what may be "true" at one moment may not be true at the next moment, thus, in order to know everything, your knowledge or Science must be dynamic and ever changing with creation over time. Therefore, there is always more to know and learn, and thus the endless cycle of creation and change in Nature. When you may think you know what is going to happen next, and use reason to try to assume the outcome, this is still speculation, because you really don't know the outcome until the event happens. When you only experience and do not attempt to reason or predict to learn, you are subject to be used, and literally have the mentality of a "victim" or "prey." As Humanity moved from being **"prey"** and **"Praying"** or calling on God (Nature) to provide, and started Creating, Controlling, and being "causes" in Nature rather than victims of the "effects," then the process of scientific discovery could proceed. Initially, modern Scientist attempted to discover Scientific "Laws," and came up with the **"Laws of Mechanics"** which were thought to be complete and infallible at that time. Further investigation into the relativistic nature of the dualities in astronomic or large scale properties of Nature and Science gave rise to the modern scientific theory known as **"The Theory of Relativity,"** that superseded concepts presented in the "Laws of Mechanics." And even more recently, examination into the constant change, creation, and "chaos" that takes place within the sub-atomic or small scale properties of Nature and Science gave rise to the modern scientific theory known as **"The Uncertainty Principle"** which is in conflict with certain concepts presented in "the **Theory of Relativity**." The analysis (separation stage in Science) of Nature has led to the discovery of "the **Fundamental Particles**" or **"Elementary Particles,"** which are the smallest particles or parts of Matter in Nature. The "analysis" or learning stage of Science has also led to the discovery of the Scientific "Forces of Nature" or "Causes" known as **Electro-Magnetism**, **The Strong Force**, **The Weak Force**, and **Gravity**. All of these forces or "causes" are "minds in Nature", and Scientist have discovered a "fundamental particle" or "brain" for each of these forces except for Gravity to date. The Electro-Magnetic, Strong, and Weak forces and their respective particles called Photons, Gluons, and Bosons all are explained by way of "The Uncertainty Principle" and other Theories in Quantum mechanics. The single theory that unites these three forces is called the **Grand Unified Theory** or **"GUT."** However, just as Scientist have not discovered a "brain" or fundamental particle for the force or "mind" called "Gravity," Scientist have also been unable to unite Gravity into a single theory that explains all four fundamental forces. The single theory that would explain all four fundamental forces is called **"The Theory of Everything"** or **"TOE."** Attempts to unify the force called gravity with the other 3 forces have given rise to the hypothetical concepts of **Dark Energy** and **Dark Matter**. Ironically, it is the force called **Gravity** which **"Unites"** or **"holds matter together"** that has been the most difficult "force" to **unify** with the other three forces. If and when the Gravity force is united with the other three forces and the "Theory of Everything" or TOE is developed, then the TOE would be able to predict the outcome of ANY experiment. At the point where the TOE is discovered, we have transitioned from the "analysis," separation, or learning phase into the synthesis, unification, and application phase in Science and Nature. It is also interesting to note that these unifying theories in science, namely the **GUT** and the **TOE** (or **GUTOE**), are similar abbreviations for one of the titles of the African deity **PTAH** which was **"Grand Architect of the Universe"** or **G.A.O.T.U.** Even TOE is similar to PTAH in pronunciation.

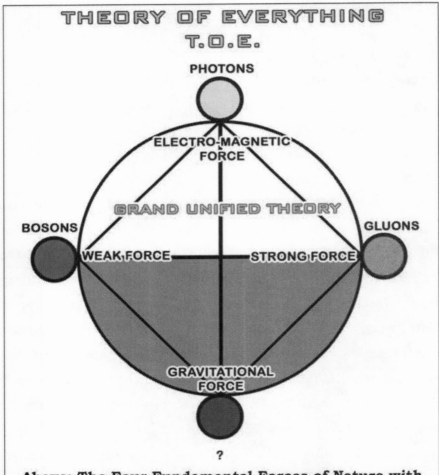

Above: The Four Fundamental Forces of Nature with their corresponding Particles. The particle for the Gravitational Force is only proven by reason and not been experienced at the time of this writing.

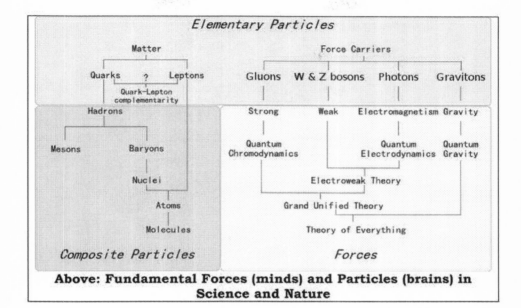

Above: Fundamental Forces (minds) and Particles (brains) in Science and Nature

The unification of information leads to comprehension and application of information. Scientists have unified the concepts of Space and Time into a "Space-Time continuum" for ease of understanding, and it would not be surprising if matter and energy are unified into a "Matter-Energy continuum" for ease of understanding; or if all of these component parts of Nature are unified to create a "Space-Matter-Energy-Time Continuum" or "Nature Continuum" which would work in conjunction with the TOE to conclude other scientific theories. This sub-section on "The Ontology of Science" has set out to answer and expound on the knowledge questions "Where, When, Beginning of, and Ending of" Science, and inform the reader about the Existence of Science as well as where we exist right now in terms of Universal Science Knowledge in Nature.

2.3. The Noölogy of Science

Following our Ontological discussion on the existence of Science, we must also discuss and answer "how" we know Science exists. The answer to the question of "How we know science exists" is a combination of both "Means" and "Method." The "Means" is the way Scientific Information is obtained, and the "Method" is the way the information is processed. The "methodology of Science" will be discussed in greater detail in the coming sections of this presentation, but the "Means" by which scientific information is obtained will be discussed in this section called "The **Noölogy** of Science." The word "**Noölogy**" comes from the Greek words "**Nous**" and "**logos**," and has a literal meaning of "the study of the mind." However, the word "Noölogy" has a broader definition referring to "systematic cognitive neuroscience," or to put it simply, "the way thoughts and information are processed in our mind by way of the **Neurons**, nervous system brain, and our senses." In the section about "The Ontology of Science," we stated that everything in existence exists in the state of either "Rational or Empirical", "Abstract or Concrete;" that is to say, in the mind as a concept, or in "reality." However, even empirical things that exist in reality are reflected in our mind as abstractions. Our nervous system and our senses are the "Means" by which information in the Empirical world is translated into the Rational world of the mind. Everything in the mind is what we "know," and thus knowledge itself is an abstraction.

An "Abstraction" is defined as a concept or idea which serves the purpose as a "general quality," characteristic, or "super category" for all subordinate concepts, and connects any related concepts as a group, field, or category. Abstractions are distinguished apart from, yet based on, concrete realities, specific objects, or actual instances. Abstractions are created by the mental process of strategic simplification and reduction of experienced information to keep only the information which is considered relevant for the purpose of the creation of the abstraction. Because of the simplification and reduction of information that takes place in the creation of abstractions, certain information is lost, ambiguous, vague, and undefined. Miscommunication (and miss-thinking) can occur due to the assumptions that exist within, or have to be made with the use of the abstraction. Words are abstractions. Even though you can use words to refer to things in the Empirical world, the words themselves are inventions from the mind. The words themselves are composed of letters, and letters are abstractions because they exist as symbolic creations of the Human mind to represent, or reflect a sound experienced in the Empirical world. The letters and words combine to form our "language" which is literally what we use to think; i.e. the **"sound of thought"**. However, language is an abstraction and an invention created by someone's mind. And so, one must ask the question, before the invention of letters, before the invention of words, before the invention of language, "What did thought sound like"?

The "sound of thought" before the invention of language (and also which led to the invention of language) was impulse or "feeling." This impulse or "feeling' is simply the information relayed to our brain from the Empirical world by way of our nervous system and senses or **sensors**. Over time, we used the sensory information to create letters, words, sounds, and other abstractions. Our senses, or sensors, provide the information input for perception, which in turn determines our awareness, consciousness, and reality. The general flow of this process of information from the external world, to the world of our mind, has 4 general steps which are depicted below:

1		**2**		**3**		**4**
Empirical World (Environment, Phenomenon, Events)	→	Information Input (Senses)	→	Rational World (processing)	→	Output (Re-action in the Empirical World)

Because everything that we know and all information we obtain is acquired by way of our senses, then it is a fact that "Reality" is determined by our perception. This also shows the "Relativity" of reality, because a person's "reality" is only as accurate as the detectable information obtained by their senses and their perception. Moreover, the use of "language" as a means to interpret sensory information skews one's perception of reality to the default assumptions, presumptions, simplifications, and reductions contained within the invention called "language."

Using Science, Mathematics, and Technology, we have invented tools to help expand the range of our senses, so we know for example, that there are frequencies of light and sound that we cannot see or hear, but are detectable using certain technology. Also, there are scents that we cannot smell, flavors that we cannot taste, and touches that we cannot feel. These technologies that increase the range of our senses, in turn expand the **"range of our reality."** Generally speaking, we Humans are considered to have "5 basic senses" that correspond to 5 sensory organs on our body connected to our brain by way of the central nervous system. The "5 basic senses" are Sight, Hearing, Taste, Smell, and Touch. The "5 basic senses" combine to create "5 acquired senses" of 1) the sense of balance, 2) the sense of direction, 3) the sense of motion, 4) the sense of time, and 5) the sense of temperature. All of the senses basically are designed to detect the general aspects of Matter and Energy in Nature; thus, all of our senses are indeed various form of "touch" in so far as some form of matter or energy must come in contact with or "touch" one or more of our sensors in order for a particular phenomenon to be "known." There are also other senses that exist within animals, and to a small degree in some humans, that enable the delectability of certain natural phenomenon that is only currently detectable by technology and instruments. Some animals and Blind humans have a sense of Echo-Location which is similar to "SONAR," which enables navigation by the use of sounds.

In Africa, an "Eye" symbol was and is used as a composite symbol to represent all of the information received to the human mind by way of the senses. Considering that the mind is divided into Two Hemispheres, the Right Eye of Horus represented information received by the Right Hemisphere of the Brain (the "Creative" or Empirical side) and the Left Eye of Horus represented information received by the Left Hemisphere of the Brain (the Analytical or Rational side).

Symbol	Description	Sense
	Nose	Smell
	Eye Pupil	Sight
	Eyebrow	Thought
	Ear	Hearing
	Tongue	Taste
	Hand	Touch
	African Symbol for All Information received via the senses by The Human Mind	

Right Hemisphere, Creative "Empirical"	**Left Hemisphere, Logical "Rational"**

The information that is received to our Brain for processing by way of our Senses is converted into electricity. The **"thoughts"** that occur in our Brain are the result of information being transmitted as **electrical signals** from **Neuron** to **Neuron**. Now that we know "what Thought is," we can actually determine **"The Speed of Thought."** Considering that "electricity" is a form of **electro-magnetic radiation**, and the fastest speed of electro-magnetic radiation in a **vacuum** (empty volume of space) is **299,792,458 meters per second**, we can then calculate the upper limit of "the speed of thought." A "Refractive Index" is the measurement of the speed which electro-magnetic radiation can travel through a particular substance. The Refractive index of Glycerol, Polycarbonate, and human skin is in the range of 1.4 to 1.6, so we will estimate the refractive index of Neuron cells as well as other Brain matter to be about **1.5**. Thus, we can calculate the fastest possible speed that electro-magnetic radiation can travel in our Brain (the Fastest possible speed of Thought) to be **199,861,639 meters per second** which is 33% less than the speed of electro-magnetic radiation in a vacuum. Also, we must state that modern scientist inappropriately call "the speed of electro-magnetic radiation in a vacuum" (which is 299,792,458 m/s) by the term **"the Speed of Light"**, however, this value is actually **"The Fastest Speed of Dark"** and the speed of **"Light"** or **visible electro-magnetic radiation** can be calculated using the refractive index of air which is 1.000277. Therefore, the speed of **"Light"** or **visible electro-magnetic**

radiation is actually **299,709,438 meters per second,** which is 0.03% less than the speed of electro-magnetic radiation in a vacuum (The Fastest Speed of Dark).

The "Speed of Thought" is also calculated in **Reaction Time**, which is the amount of time it takes for a person to receive some information in the form of stimuli and respond to it. **The average Human reaction time is about 0.25 seconds**. The calculation of Reaction Time is actually the sum of the amount of time it takes to transmit some information, the amount of time it takes to process the information (think), and the amount of time it takes for the Brain to send messages to the body appendages to respond. Therefore "Reaction Time" is not a perfect indicator of "The Speed of Thought" but it gives a relatively good estimate. Considering that **the average Human brain is about 980 cm³**, then, the average speed of thought would be just under **0.4 meters per second**. The calculation of "Reaction Time" is like the calculation of "Information Round Trip Time" or a Ping to test the speed of a computer. Comparing the Human Brain to a computer processor, the Human Brain is capable of processing 100 million Instructions per second, and therefore **the Human Brain would be equivalent to a 17THz Pentium processor computer**. While 17THz is much faster than the fastest super-computer in the world at the time of this writing, the 17THz frequency places the frequency of the Human Brain in the **Infra-Red range** (1 to 430 THz) of the Electro-magnetic spectrum.

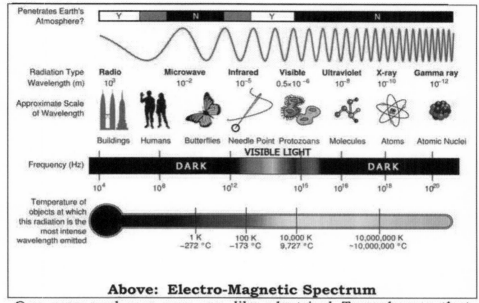

Above: Electro-Magnetic Spectrum

Our eyes and our ears are like electrical Transducers that convert light and sound into electricity. The electrical signals from Transducers are then processed by a **silicon** computer chip called **micro-processors** in computers. Much like a Photo-detector sensor converts light into electricity; our eyes also convert light into electricity to be processed by our Brain. Much like a Microphone converts sound into electricity; our ears also convert sound into electricity to be processed by our Brain. Therefore, our Brain is similar to the processor of a computer. The silicon devices that are used in computers today were preceded by **vacuum tubes** in earlier times. Much like a vacuum tube, the pressure inside of our head is also less than atmospheric pressure so that the electrical signaling between Neurons can take place. Sights, sounds, smells, flavors, and feelings are all stored in our brain as electrical signals. It is a myth that we only use 10% of our Brain.

While humans can detect a certain range of Electro-Magnetic Radiation called "Light" by way of our vision, some animals have a sense of Electro-Reception (the ability to detect electric fields) and Magneto-Reception (the ability to detect magnetic fields). Also, humans can detect a certain range of temperatures, and temperature is related to "Pressure" by way of the "Ideal Gas Law" $pV=nRT$ (where p and T represent pressure and temperature in the equation respectively), however, some animals have a sense that enables them to directly detect pressure. The information or knowledge obtained by Human beings by way of the senses can be thought of as **"common sense."** It just so happens that "common sense" or "sense" and "sensibility" in general, is also associated with something "reasonable." Often times people say something "makes sense" when they mean that the information is "reasonable." Of course, "Reason" is actually what processes the information obtained by the senses. Reasoning occurs in step 3 in our information flow chart, after the input of information to the brain via the senses in step 2. However, the reasoning process does in fact "make" sense or senses; and the senses made by reason have been categorized into an area of paranormal ability and parapsychology called **"Extra-Sensory Perception"** or **ESP**. Considering that the definition of "reason" is "a mental faculty that is able to generate conclusions and explanations from processing certain given information," then examining the function of the ESP psychic senses shows how they are generated by reason.

The "extra-sensory perceptions" or those senses "made by reason" are Psychometry, Intuition, and Clairvoyance. **Psychometry** is defined as the ability to gain information about an object by touch, observation, and experience, and then being able to reasonably induce relevant associations about the object or the object's owner from the given information. **Intuition** is defined as a quick for of reasoning where thoughts, ideas, conclusions, and insightful explanations come quickly into the mind without much reflection, and is also closely associated to **Emotion**. Clairvoyance is most commonly defined as the ability to use inductive reasoning to gain information about persons, places, things, and events that have occurred in the past or which will occur in the future from some given information in the present. Consider for a moment the ontological status of "Time." The three degrees of time are the past, the present, and the future. While past, present, and future are all abstractions, the only aspect of Time that you can ever prove empirically is the Present or **NOW**; all other aspects of time (past and future) are only abstract rational concepts proven by reason. You cannot ever empirically show that the past or the future exists, but you can use reasons to show "REFLECTIONS OF YESTERDAY" and "PROJECTIONS OF TOMORROW" which will all take place TODAY. And so, all information about the past and future, including Clairvoyant abilities are forms of inductive reason that enable the individual to "re-trace the past and foretell the future" in the present.

To the degree that these conclusions generated by reason that are collectively called "extra-sensory perceptions" or ESP are accurate, it would appear that the individual generating these conclusions are in fact "detecting" something outside of the range of the "common sense" of common people. But actually, the individual generating these conclusions is receiving the same information via the same 5 senses as everyone else; however, they have developed a better and more accurate way of reasonably processing the information to generate accurate conclusions and explanations. So to this regard, we see the abilities of Psychometry, Intuition, and Clairvoyance as not paranormal at all, but the result of proper and accurate information processing called reasoning. However, the "extra-sensory perceptions" called Remote Viewing, Channeling, Mediums, Séances, Telepathy, and Telekinesis do not appear to be senses at all, but rather extra-ordinary abilities. In the case of these abilities which deal with some form of communicating information by way of some unseen and unknown method, or controlling and moving object by some unknown method, these abilities would indeed require a "sensor" that is not common or active in most people. Perhaps these abilities are tapping into some form of energy field like the Electro-Reception, Magneto-Reception, and Pressure Reception senses in some animals. In the case of telepathy, it would require at least two people with similar abilities to transmit and receive information. Therefore, we can conceive a situation where "Telepathy" could be considered a "sense" or ability made by reason.

Considering sign language, which is the ability to communicate by way of gestures, if two or more people were able to devise a communication system consisting of very subtle eye movements, facial expressions, and subtle gestures, then the two individuals could be able to communicate with one another in a room and it would appear that nothing is going on and nothing is being "said" to an external observer who did not know the language. In this regard, Telepathy could be considered a form of "reasoning" to interpret the subtle clues offered by another individual. However, in the proposed scenario, the "telepath" would have to receive some form of information from another individual for this to work, and this method would not work over long distances where no information is transmitted or received by the two individuals; unless they are able to send and detect information by way of some "energy" field.

As we stated earlier, all information that we know is obtained by way of our senses, the information must make contact with, or be "felt" by our sensory organs in order for it to be detectable. So our senses are in fact related to what we "feel" or our "feelings." Also, the word "feelings" is associated with "Emotions." Emotions, like certain extra-sensory perceptions, are the result of reasoning. Emotion is the result of a quick processing of information by way of reason much like "Intuition." It is not surprising that "Intuition" and "Emotion" are generally associated with Women and "sensitivity".

There are 8 basic emotions that can combine to form other more complex emotions. The 8 basic emotions result from an individual obtaining some information by way of the senses, and then quickly processing the information by way of reason. There are three basic premises that are used by reason to create the 8 basic emotions. The three basic premises are: 1) Is the Information received expected to be Positive or Negative for me, 2) Is the Information received something that I can control or something that I cannot control, and 3) Is the information something that has happened in the past or will it happen in the future. All three of these categories deal with inductive reasoning. The quick "yes or no" evaluation by way of inductive reason in an individuals mind to these three categories gives birth to the 8 basic emotions listed below:

Emotion	EXPECTATION 0 = Negative 1 = Positive	CONTROL 0 = Not in Control 1= In Control	EVENT 0 = FUTURE 1 = PAST
Fear	0	0	0
Sadness	0	0	1
Disgust	0	1	0
Anger	0	1	1
Anticipation	1	0	0
Surprise	1	0	1
Trust	1	1	0
Joy	1	1	1

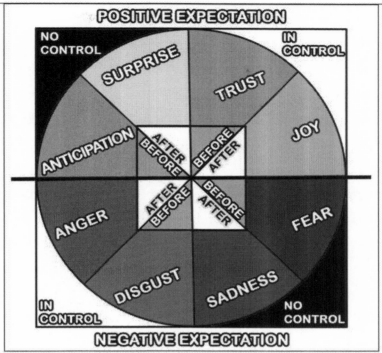

The evaluation of information as being either positive or negative for the individual, and whether or not the event has happened or will happen are inductive reasoning about Space (environment) and Time. The evaluation about whether or not the person can control the information or event is the result of inductive reasoning about one's self, "sense of self," or "self-esteem." The dichotomy in self-esteem manifests in many forms such as "introverts and extroverts" and "passive and aggressive" personalities. Since emotion is the result of reason, we recognize that animals do in fact have the ability to reason because they show emotion. If an animal runs away from someone or something in fear, it is the result of information received and processed using inductive reasoning to come to the conclusion that there was danger.

The quick form of inductive reasoning which is called "emotion" can lead to poor reasoning or things not being thought through completely; this is why it seems that emotion and reason are at odds, but in fact emotion is based in reason. It is important to reason through information and experiences and not be quick to make a judgment because the information experienced may be an illusion, delusion, deception, or false perception. However, some response is better than no response. The responsiveness to stimuli or information is what determines if someone is conscious. **Unconsciousness** is defined as the lack of responsiveness or reaction to information. Unconsciousness is a characteristic of a **comatose** state and **sleep**, which are both similar to **death**. Thus, responsiveness and reaction are necessary for life. However, just because someone is nonresponsive does not mean they are unconscious, in a coma, asleep, or dead; it could be that there is a problem with their "information processor" called reason. The physical food you eat is processed in your digestive system and energy is extracted from the food to sustain you, and the waste is expelled. The information that you receive by way of your senses to your brain also goes through a similar processing cycle as food. The information that you experience makes up your "conscious mind" (that which you are aware of) and is ready to be used for decision making and daily activities. Information that you experience (eat) but are not aware of and don't use, goes to the mental **"waste land"** called the **"unconscious mind."**

The **Unconscious mind** is also sometimes called the **Sub-Conscious mind** and is the part of the mind that collects all the information and mental phenomena received by the senses that the individual is not "aware" of, and is not part of the "conscious mind." The "unconscious mind" or "sub-conscious mind" is the source of "Dreams," which are a means to excrete or get rid of un-used information and mental phenomena. The "unconscious mind" or "sub-conscious mind" also stores information on things we have learned so well that we do them without thinking. For example, your heart beat is controlled by your unconscious mind because it beats without you thinking about it or controlling it. The "unconscious mind" or "sub-conscious mind" also stores information that may seem forgotten, but may be used at a later point in time; this is said to be the source of unconscious "insight," **"inspirations,"** and **"revelations."** What is called the "the **Super Conscious**" or "**Higher Conscious**" is also called the "god consciousness," "Buddha consciousness," or "Christ consciousness" in various Indian, Asian and "New Age" Philosophies and beliefs. The "Super Conscious" denotes a type of "consciousness" that is supposedly higher, superior, or greater than that of other humans enabling the individual with the ability to know or be aware of "reality" or "god" to a greater accuracy than other people. Of course, following our discussion in "The Preface of Science" we know that the method proposed by these various philosophies is African in origin, and is the S.O.S.A.S.I.S. method of processing information.

Moreover, the purpose of information is to be used for activity; thus, what good is a "Super Consciousness" without "Super Activeness." It is the reaction and response to information that determines if you are "alive" and shapes your reality and existence.

Existence is the state of existing, "being", or "to be." The awareness of this state of "existence" or "reality" is defined in the Mind by way of the senses. Since reality and existence take place in the mind via the senses, reality can be replicated. Visual reality can be replicated using holograms and 3-dimensional images, sound reality can be replicated via various recordings and simulations, smell reality, taste reality, and touch reality all can be artificially replicated by sending and stimulating the correct electrical impulses to the individual's brain. Therefore, the entire experience of "Reality" can be simulated, either for the purpose of experiment or deception. If your entire reality is simulated, meaning someone or something has contrived and created all of your sensory information, then how would you ever know it is a simulation? If you have reason to believe that it is a simulation, then the question would be "where or what was the sensory information that you received into your mind to make you think that your reality is a simulation?" Reason is the process by which information is deemed correct (True) or incorrect. Reason uses evidence in the form of experience to determine validity of cause and effect assumptions.

When what we reason to be true and right does not match up with our reality, we often abandon the reason or thought for the experience of reality, and assume our reason must be wrong. At best, our reason helps us to understand our reality. If all of the sensory input to our brain was ever simulated, there may be no way we would know we were living in a "virtual reality." It is possible that our reason may allow us to discover the deception, but more likely our reason would begin to help us learn, understand, and "make sense" of the "virtual reality" in which we were living.

Statements have been made such as "Seeing is believing" or "Hearing is Believing and Seeing is Knowing". In either case, the information acquired via the sight sensors or sound sensors is information, and any "cause or effect" explanation or description of the observation produced by reason would have to be checked out by The Science of Sciences to determine if the explanation or description is Right Knowledge (correct, aligned with reality). If the observation information is artificially simulated for the purpose of deception, it still is a valid observation, but the explanations and implications based on the observation are likely to be skewed. The question is then asked, "How do we know we are not in a simulated or 'virtual' reality right now?"

There are many instances where "scientific" explanations defy "intuition" and reason. The initial reasoned hypothesis or belief does not always have to line up with the evidence experienced. In the cases where the hypothesis or reason does not align to experience, we change our hypothesis or reasons. However, if the evidence and the experience defy reasoning, then a serious question arises: What is real and what is Right? Given a collection of observations, the observations that seem to be outliers, or somewhat different from the other observations, usually have different cause and effect explanations for given the conditions. This is what separates an "Illusion" from a "virtual reality." In the case of an Illusion, or any information detected by the senses, there may be "more than meets the eyes." Illusions are experiences that defy reason, and they defy reason because they unlike any other similar observations in our reality, and are not replicable by way of science each time the conditions are the same. However, given the same observation that was considered an "illusion" in one reality, if it is common place in a "virtual reality" we would not consider it an illusion at all. This is also true for phenomenon considered **paranormal activity**. Paranormal events often defy reason, and are not scientifically explainable or replicable. However, given a reality or virtual reality where the phenomenon that we consider "paranormal" in this reality were common place, then we would not consider them paranormal at all, but rather regular events. Thus, deception is only detectable when it out of place with reality.

Reception of information requires perspective, thus there is a limit to how much information can ever be obtained, for one can never fully observe one's self. The square is the finite, the limited one, the experiencing one. The circle is the infinite, the unlimited one, the empirical one, the one providing the experience (information) to the square. These dualities alternate in their perspective. You can study your surroundings, but you cannot fully study or observe yourself without destroying yourself. Information is the Natural Resource that is needed by all beings, and thus the essence of existence. Information is the foundation of Intelligence, Understanding, learning, survival, life, and existence; for how would you know you were alive or how would you know you exist without information? It is by way of "The Noölogy of Science" that we can reason one of the **"Meanings (or purpose, or reason) of Life."** In architectural science, there is a term called **"Form follows function"** which basically is a phrase to say that the shape or form of something is indicative of the function or purpose of the thing. Thus, we can say that since we have senses to acquire information, and a brain to process information, then based on our "Form," our function, purpose, and reason is to acquire information and process information, and learn. This section called "The **Noölogy** of Science" presents mental information acquisition and processing system. To determine if knowledge is right, or if information is correct, or if data is accurate, you first have to have input of information to the mind by way of Noölogy.

2.4. The Epistemology of Science

We conclude this discussion on "The Introduction of Science" which establishes our Philosophy of Science by now discussing the "Epistemology of Science." The word "Epistemology" comes from the Greek word epistēmē, meaning *"knowledge, science."* Epistemology is a branch of philosophy that establishes a "theory of knowledge" (theory of science) by investigating the nature, methods, origin, limits, and definitions of knowledge, and truth. Considering that we have covered most of the epistemological facets of Science in the previous sections, we will use this section entitled "The Epistemology of Science" to discuss the limits and definitions of knowledge. We also will discuss the semantics of our definitions of certain terms related to "knowledge" such as truth, fact, belief, faith, feel, think, understanding, comprehension, justification, trust, confidence, certainty, skepticism, doubt, and proof.

We define "Knowledge" as all information that is "Known" to an individual or group. Knowledge can be true or false, right or wrong, accurate or inaccurate, correct or incorrect, and these classifications of knowledge are determined using logic and reason as proof. It is possible to "know" some information that is completely wrong, however since the information is known, regardless of the accuracy of the information; it is possible for that wrong information to still be considered "knowledge."

Thus, knowledge is equivalent to information that is "known." Information that is "known" to an individual or group is acquired by way of sensory experience. Experience can be either "First hand experience" or "Second hand experience" (which will be discussed in more detail in the section entitled "The Experiences of Science"). When information is experienced by way of the senses and stored in the **memory** of the mind, the information can then be classified as "known," and information that does not fit into the "known" category is considered "unknown." Of course, the "Memory" is defined as the ability to store or retain information and sub-divides into what is called the **"Short-term Memory"** and **"Long Term Memory."** The short-term memory is also related to what Psychologists call the **"Sensory Memory"** which briefly retains information detected by the senses, and the **"Working Memory"** which is used for immediate processes of reasoning, comprehension, and learning in the mind. The short-term and long-term memory works in conjunction with the conscious and unconscious (sub-conscious) mind. The combination of the "Known" and "Unknown" dichotomy of Information classification breaks down into four categories:

1. **Known-Knowns**

2. **Known-Unknowns**

3. **Unknown-Knowns**

4. **Unknown-Unknowns**

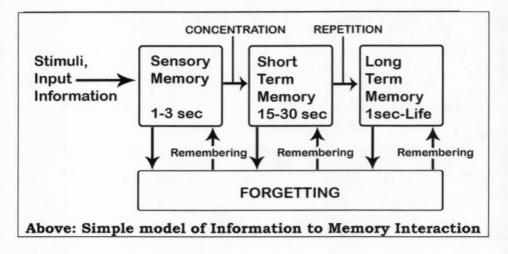

Above: Simple model of Information to Memory Interaction

We define the **"Known-Knowns"** category of knowing and knowledge as that information which has been experienced by the individual, of which the individual is aware, and is available in the conscious mind to be retrieved from the memory and used at will. We define **"Known-Unknowns"** as information that the individual has not experienced, but is curios about and has reason to believe exists based on the information that has been obtained by way of experience. We define the **"Unknown-Knowns"** category as information that has been experienced, but either forgotten or not vividly remembered. The "Unknown-Knowns" category of information also includes information in the unconscious mind and also unknown connections and unknown relationships between known information that may not be realized. We define **"Unknown-Unknowns"** as information that has not been experienced and that we have no reason to believe exists. Information in the "Unknown-Unknowns" category has not been detected and is completely outside of our intellect.

In earlier chapters we spoke about how the Ancient Egyptian concept called Sia or Saa was likened to "the mind" and intellect. We would also like to state that the concept called "Nu" was considered "chaos" and a state of being indefinite and undefined in Ancient Egypt and the concept called "Ta" would be considered a state of defined form. Also, in Ancient Egypt, the word "Pa" or "Pu" served as the "definite article." Thus, our concepts that we defined as Known-Knowns, Known-Unknowns, Unknown-Knowns, and Unknown-Unknowns could have been considered **"Pa Ta," "Nu Pu," "Nu Saa,"** and **"NuNu"** respectively. It is our belief that "Nu Pu" which represents what we call **Known-Unknowns** is like saying "I know that I do not know," "I know this is Vague," or "I am certain that I am Uncertain." The concept "Nu Pu" contains two words that are opposites in terms of their definitions as it relates to being definite. The category of **Known-Unknowns** for example would be like someone telling you "I have a secret," because now you have experienced information that lets you know that there is something that you do not know. This category of information is **the most spellbinding** and captivating category of information to the curious mind if there is not a proper method of transforming the "unknowns" into "knowns". Presenting information in the "Known-Unkowns" form is the **"Science of Spellbinding"**. Transforming the "unknowns" into "knowns" is Liberation and also called "solving for x" in the mathematics of Algebra. This category of information is also what leads to new discoveries, new inventions, and new creations if one can properly transform the "unknowns" into "knowns."

The concept called "Pa Ta" which is similar to what we call "Known-Knowns" could also be called "Saa Pu" as these concepts deal with being clear and definitive in mind. It is our belief that "Nu Saa" which represents a "Chaotic Mind" or "undefined mind" which we call **Unknown- Knowns**, gave rise to the Greek word **Nous** which refers to "mind" or "intellect." Information that is considered **"Forgotten"** is information that has be experienced and has been in the Known-Knowns category and has passed into the Unknown-Knowns category and the sub-conscious or unconscious mind. Information can be Remembered, Recalled, Recollected, or Realized in the Known-Knowns category and the Unknown-Knowns category. Information in the Known-Unknowns category can be sought after, pursued, Discovered, Uncovered, Revealed, and Learned.

Information classified in the "Known-Knowns" category or the "Known-Unknowns" category can be further classified and qualified as **"True."** Synonyms to "True" or "Truth" are accurate, correct, and valid with the corresponding antonyms being False, Inaccurate, Incorrect, and Invalid. We define Truth (and the synonyms related to truth) as a means to measure or determine by way of reason how well something Abstract (Rational) reflects and represents something Concrete (Empirical). The relationship between "Truth" and "Reasonable Thought" was depicted in Ancient Egypt with the Goddess Ma'at (Truth) being United in Nuptials as the wife of the deity Tehuti (Thought).

Also, "Truth" as a measurement between the Empirical and Rational worlds was depicted in the famous scene from the **"Papyrus of Ani"** where "truth" is symbolically weighed, measured, or determined by way of "reasonable thought" (Tehuti) on the Scales of Ma'at, which would permit the passage to the Over-world (the Empirical world) or stay within the Underworld (the Rational world).

Above: Scene from the Papyrus of Ani where "Truth" was the "balance" between the Empirical and Rational worlds determined by Reasonable Thought.

Since "Truth" is a determination between two things, then there must be at least two things in order to determine "Truth". If there is only one thing, then "Truth" is a given, and as one divides into two or more things, then Truth must be determined. Thus, the Origin of Truth was the "Big Bang" separation when everything in existence separated from a singularity into a multiplicity. Also, as everything in existence is expanding, chaos or entropy is increasing, and thus "Truth" and "Order" (Ma'at) is decreasing.

Information that is not "True" or "False" is **"Unverified."** Complete separation is the maximization of Falsehood and Chaos and the minimization of Truth and Order, whereas complete unification is the maximization of Truth and Order and the minimization of Falsehood and Chaos. As long as there are at least two, there will be some degree of falsehood. Separation enables Nature to observe it self, but the more separation, the more falsehood. Unification enables the reduction of falsehood between the Unified entities, but does not enable complete observational study of the unified entities; for one cannot analyze (or break down) itself without becoming two. Because Truth is determined between at least two things, then Truth is also a means to measure Understanding and Comprehension between two things. When there is acceptance and agreement in the Comprehension of the Truth between two things, then unification can occur. Falsehood can be intentional or unintentional. When falsehood is unintentional, then the misunderstanding of the "Truth" between the Observer and that which is being observed is a matter of perception or point of view. When falsehood is intentional, it is actually a creation, technology, or application of knowledge invented for the purpose of obtaining some desired outcome of the one transmitting information to the receiver. Falsehood can come in the form of "Noise" or "Quiet"; that is to say too much information or too little information. Noise must be filtered by way of reason to determine truth, and Quiet must be extrapolated by way of reason to determine truth.

Truth is determined by mental logic and reasoning. Thus "Truth" requires certain premises and assumptions. Therefore, the "Truth of Truth" or the accuracy of truth is dependant upon the truth or accuracy of the premises or assumptions. When one determines or judges or accepts something as "Truth" they have done so based on prior judgments or assumptions that all the previous premises of information presented are "True". For example, consider the two premises and the following logical conclusion:

- Premise 1: If   A = B

- Premise 2: and   B = C

- Conclusion: then it is TRUE that A = C

The "Truth" in the above conclusion is only true if A = B. If for whatever reason the premise or assumption that A = B is not true, accurate, valid, and correct, then the conclusion will not follow from the premise. Thus, the "Truth" determination requires **Faith**, **Belief**, and acceptance that the premises are "true." In the above example the Reason and Logic is Valid. **Logical Validity** is synonymous with **Logical Truth** as long as the conclusion follows from the premises. However, Reasoning can be Logically Valid or Logically True, with completely false premises. **Sound Reasoning** is when the Reasoning is Logically True and Logically Valid, and all of the premises are "True," and thus Sound Reasoning is the "Truth of Truth" or the highest form of accuracy and validity in information that can be proven or justified.

A "True" premise is a determination of how accurately the definition of an abstraction matches the intended empirical thing it is supposed to represent. A "True" premise is also called a "**Fact**." The Truth about Facts is that All Facts are "True" (if properly stated). Facts are statements of things Empirical and Concrete using Abstractions called "words" which express actual events. Someone stating a FACT does an internal logic and reasoning assessment to determine that the Abstract words used to describe an empirical/concrete event are "True." Thus, the assumption of "Faith" that occurs in a statement of FACT is that the definition of the words being used to describe an event is indeed accurate. Statements of "Fact" contain relative "truth" as determined by the observers based on the prior information or prior premises that they accepted as "true." Facts in the form of "relative truths" are reconciled by way of reason and semantic definition of the words used. True Logic is a determination of the accuracy of the rational method of processing information. "Invalid" or "Untrue" logic is reasoning that does not follow from the given premise information. **Sound Reasoning** is the combination of **True premises** and **True Logic**. The process of presenting the Logic and Reasoning of information processing is called **Proof** and **Justification**. Proof is the act of presenting the reasoning behind rational abstractions and Justification is the act of presenting the reasoning behind empirical actions. Proof occurs before an action and Justification occurs after an action.

Both Proof and Justification use logic and reason to explain why something is or was believed or accepted to be true. Justification uses empirical evidence along with logic and reason to provide explanations for conclusions. People justify withholding information stating cliché things such as "you are not ready for it" or "I don't want to cast pearls to swine", but in fact nature has placed a mechanism within beings so that if a person is indeed not ready for some information, they would not be able to receive it (via the senses) or they would not be able to understand (or make sense of it). We also specify the difference between "Proof" and "Justification" in that we consider "Proof" similar to what is called "*a priori knowledge*" or Conclusions proved based on logic and reason and independent of evidence and experience, whereas "Justification" is similar to what is called "*a posteriori knowledge*" or information justified using experience and empirical evidence. We define "**Belief**" as the acceptance of Information as "True." We consider **Faith**, **Trust**, **Certainty**, and **Confidence** as related to belief in that all of these words deal with the acceptance of some form of information as being "true." Of course belief, faith, trust, certainty, and confidence can be right or wrong; they are right when what is accepted as true is actually true, and they are wrong when what is accepted as true is actually false. The opposites of these terms which do not accept information as "true", namely **Doubt**, **Skepticism**, **Distrust**, **Uncertainty**, and **Cynicism** can be right or wrong also.

Everything that someone "Believes" or accepts as "true" is known and thus part of knowledge. It is possible to know some information that you do not believe or accept as "true"; But, it is not possible to believe something that you do not know. Truth can exist without you knowing it, and thus you would be unable to accept and believe it. It is also possible to know some information is true and accurate but not accept or believe it. People do not accept that some information is "True" because they either disagree with the logic and reason of the Proof of the truth, or they disagree with the evidence used as Justification of the truth. When Sound Right Reasoning is presented with True Premises, True Evidence, and Valid Reasoning, this is the highest form of "Truth" in information, and any one who rejects it is indeed **delusional**. Skeptics ask if it is possible to "know" anything, however, based on out definition of "knowledge" which is a body of information that is "known" it is indeed possible to "know" something by virtue of the fact that you are able to ask the question you must know how to think, how to speak, and what to say. In examining the Epistemological "Limits of Knowledge" we now ask the question "Is it possible to know everything?" Consider that the fastest thing in existence is electro-magnetic radiation (also called "Light") this would also give us the smallest unit of time; Consider that the entire universe is expanding at an increasing rate, this would give us the largest unit of time and space; Consider the smallest thing in existence are the "fundamental particles of nature" and the largest thing in existence is what is called the **"Observable Universe"** (that is everything in the

Universe that can be experienced), then this gives us our limits on all of Space, Matter, Energy, and Time which are the component parts of all of Nature. Since there are limits on all of nature and the "fundamental particles of nature" are quantized (meaning able to be counted), then these limits on nature would also be the limits on all of the information in Nature, and thus the limits of all Knowledge.

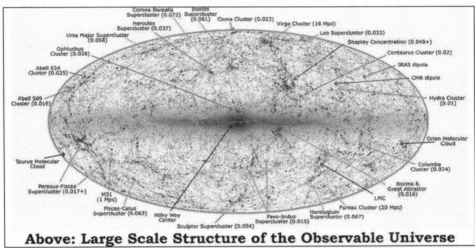

Above: Large Scale Structure of the Observable Universe

Our "Short-Term Memory" has a limit of about 7±2 elements (as few as 5 and as many as **9** elements), but our "Long Term Memory" is able to store an unlimited amount of information for an indefinite amount of time. Thus, it may be possible to Know Everything and be **Omni-Science** if you can be **Omnipresent** and experience everything, and the application of "knowing everything" would enable the ability to be "All Powerful" or **Omni-Potent**; these are characteristics that religionists attribute to "God," and thus it is apparent why "The **Science of Sciences and The Science in Sciences**" is also called "**The Science of God**."

3.0. THE SCIENCE OF SCIENCES

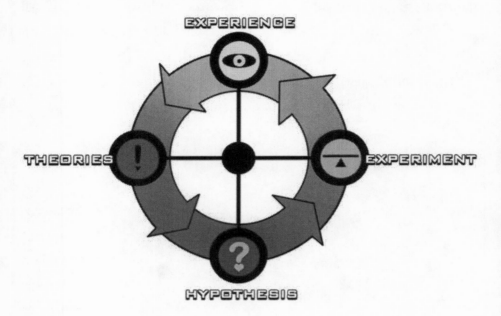

Now that our "Philosophy of Science" has been established which defines important definitions of concepts and terms, we will now discuss the Methodology of Science, which will answer the knowledge question of "How," and provide guidance, directions, a Path, and a "way" to practice Science in order to obtain knowledge of Nature, reality, and "God." As we mentioned earlier, the Methodology of Science consists of the Empirical and Rational dichotomy or duality. This section discusses "The Science of Sciences" which would also be known as "The Scientific Method" or the "Empirical Method" in modern terms. The figure above shows the four main points of interest when we discuss "The Science of Sciences" which are namely: Hypothesis, Experiment, Experience, and Theory.

In books written by the author Amunnub-Reakh-Ptah who was mentioned in "The Preface of Science" section, these four main points of the "Science of Sciences" method, namely Hypothesis, Experiment, Experience, and Theory, are similar to what he calls **Wu-Nuwaupu**, **Wu-Nupu**, **Asu-Nupu**, and **Naba Nupu** respectively in the Nuwaupian philosophical and cultural system. In books written by the author Afroo Oonoo who was mentioned in "The Preface of Science" section, the four main points of the "Science of Sciences" method are similar to what he identifies as **Religion**, **Noone**, **Noponoone**, and **Pantheism** respectively. The four main points of the "Science of Sciences" method are also similar to the four positions of the sun as described in the African Congo Dikenga cosomogram, namely **Masoni**, **Kala**, **Tukula**, and **Luvemba** respectively, and also to the four positions on the Opon Tray used in West African **Ifá** (or **Nupe**) system.

Evidence, Experience, and Reason are a "triad" in "The Science of Sciences" which may be distinguishable, but are inseparable. To say you have "evidence" without experience and reason is untrue because "Evidence" requires you to "experience" the evidence, then you would need "reason" to know it is evidence, and reason to draw conclusions from the evidence. To say you have "reason" without experience and evidence is untrue, because you have to experience reason, plus reason needs something to reason about, and this something comes in the form of evidence.

To say you have experience without evidence and reason is false because all experiences provide evidence which is then used to reason, plus all evidence experienced happens for a reason. Thus, you **Experience Evidence and use it to Reason**. All three are always practical. You cannot say you have ever used one without having simultaneously used the other two. In the "Science of Sciences" Evidence is obtained by way of Experimentation, and Reason is what generates Hypotheses. In the "Science of Sciences", Theory is a form of "Empirical conclusion" that connects and transforms Experience into Reason (Hypotheses) giving us our **Quadity**, or "four positions of the sun."

We depict the "Science of Sciences" geometrically using the Circular shape because the circle represents the Sun, the infinite, and continuous which is associated with Empiricism as discussed in the "Preface of Science" section. For the purpose of comprehension we separate the Methodology of Science into "The Science of Sciences" and "The Science in Sciences" which are Empiricism and Rationalism respectively, however, the Empirical method of "The Science of Sciences" has Rational components, and the Rational method of "The Science in Sciences" has Empirical components. This method of combining Empiricism and Rationalism is called **Neo-Positivism** and also **"Natural Theology"** in modern philosophy.

The purpose of the S.O.S.A.S.I.S. method is to process information (Reason Soundly) so that correct, right, valid, accurate, and true conclusions can be reached about the information which in turn leads to right, correct, efficient, and productive actions and activities. The process of transforming the unknown into the know is the mental movement through **"3 stages of unknown"** called **"triple darkness"** into the stage of "known" called "light." However, this movement is not linear but rather cyclic because at the point that information is "known", the mind begins to speculate about the implications, causes, and effects of the information; and this is the process of creation.

From Triple Darkness into Light		
Known Knowns	**Light**	
Known Unknowns	**Dark**	↑
Unknown Knowns	**Darker**	
Unknown Unknowns	**Darkest**	

Where as most people think of "The Scientific Method" as something that only pertains to the field of "Science" **we extend these concepts of the Scientific Method to all mental processes that occur in life**. In clarifying the four major points of the "Science of Sciences" method also called "The Scientific Method" and "The Empirical Method", we will start with the "Darkest" position of the "four positions of Sun" in the empirical "Science of Sciences" method which is stage of mental speculation called "Hypothesis."

3.1. The Hypothesis of Science

HYPOTHESIS

The Hypothesis stage in "The Science of Sciences" is the point when explanations are generated for experienced phenomenon. The Hypotheses of Science are the **assumptions**, **guesses**, **speculations**, and **beliefs** of Science (and in Life). The word "Hypothesis" comes from the Greek word "thesis" meaning "to put" and the Greek word "Hypo" meaning "under"; thus the etymology of the word "Hypothesis" is **"to put under."** The Hypothesis of Science is much like a **deep ocean abyss** in which you can drown and be taken under if the hypothetical speculations are unreasonable. The Hypothesis stage is a pivotal stage in "The Science of Sciences" because it can determine the success or failure of the practitioner. While "The Science of Sciences" method is Empirical, the Hypothesis part is Rational; meaning it takes place in the mind. The assumptions, guesses, speculations and beliefs called "Hypothesis" are generated rationally by way of logic and reason on empirical information. Due to the speculative and inquisitive nature of Hypothesis, we use a **"Question Mark"** symbol with a **"Heart"** as our symbol for Hypothesis in "The Science of Sciences." In ancient Egypt, the heart was called **"Sia"** or **"Saa"** and was considered to **"guard"** the **mind,** or in some cases to actually be the mind. The heart represents **intuition** and **feeling** which are indeed a part of developing Hypotheses, however, as we discussed earlier, intuition and feeling are various forms of **"Reason."**

The development of Hypotheses as explanations for experiences is a form of reasoning called **Abductive Reasoning**. The word "Abduct" means **"to carry off"** and it is very true that the speculating and "guessing" that takes place in Hypotheses can **"get carried away"** and **"go overboard"** at which time the explanatory power of the Hypothesis diminishes and subsequently the productive power of the individual diminishes also. When a set of possible explanations for an experience is derived, one must avoid irrelevant information that cannot be proven to be a contributor to the experience. **The assumptions and beliefs called Hypothesis are unavoidable in life**. However, at that point in time when a Hypothesis must be generated, the Best explanation can be determined using **Reason**, **Simplicity**, and the **Mathematics of Speculation** called **Probability** and **Statistics**. Statistics and Probability enable the individual to calculate which explanation, is most likely to **Happen**. Hypotheses are also related to the Antecedent of a proposition in a logical statement. The Hypothesis stage of the "Science of Sciences" method is the point where **Inspirations** and **Revelations** lead to **Explanations**. Inspirations which lead to Hypothesis Explanations usually come from outside ones self by experiencing some related event that inspired the formulation of a Hypothesis Explanation. Revelations which lead to Hypothesis Explanations usually come from inside ones self by way of reasoning and making connections between prior experienced information. Both Inspirations and Revelations are **Motivations** for Hypothesis Explanations. Hypotheses are constructed as statements of **Expectations**, **Hopes**, and **Desires**. The etymology of the word "Hope" is related to the word **"Hop"** meaning **"jump,"** and it is very true that individuals can form Hypothesis that "Hop" or "jump" to unreasonable conclusions. It is said that Buddha said **"Desire [Hypothesis] is the root of all suffering"** and this is very true especially if the individual chooses to Desire, Expect, or Hypothesize about something that is unreasonable, unrealistic, and unattainable. Desire or Hypothesis is only the root of all suffering when it Fails or is found untrue. As the old saying goes "Hope [Hypothesis] springs eternal," meaning people always want to be optimistic, however having unrealistic Hopes (Hypothesis) will surely dampen that optimism. Modern Scientists require that Hypothesis (desires, expectations, beliefs, guesses, speculations, etc.) be **Testable**, **Confirmable**, or **Falsifiable**. Hypothesis must also be **Practical** and **Reproducible**. In the Christian Bible in Hebrews 11:1 it states "Faith is the substance of things hoped for, and the evidence of things not seen." While we agree that "Faith" or Hypothesis is the substance of things hoped for or desired or expected, we by no means advocate that "Faith" or Hypothesis is "evidence" for things not seen; that is Delusion! Someone who holds on to a Hypothesis, faith, or belief after it has been proven wrong will likely have many problems and experience much suffering in life. The "Science of Sciences and The Science in Sciences" is Liberation information that enables the individual to solve life's problems, and **"firmly believing"** something that has been proven wrong is extremely detrimental. It is impossible to "firmly believe" anything, because "beliefs," faith, assumptions, speculations, and Hypotheses are not firm, but rather, shaky and subjective. You have the power to create! If there is something that you "firmly believe" that has indeed been proven wrong, in order for you to not be delusional in your acceptance of a wrong belief, you can bring your Hypothesis or Belief into reality by creating it. In order to not be insane, you must bring the Hypothesis and Beliefs of your Imagination into reality for everyone to experience rather than living in your own imaginary reality in your mind! If your Hypothesis fails but for what ever reason you truly desire for that Hypothesis to be true, then you must create a circumstance or scenario where that hypothesis can be true. The Concrete (Empirical world) is the base and foundation of The Abstract (Rational world), and the Abstract is the motivator of the Concrete. A **"Working Hypothesis"** is an assumption or speculation that has been tested, proven, and accepted as an accurate explanation because there are no better alternatives available. Over time, as new information is acquired by way of experience, hypothesis (beliefs) must be modified and changed to reflect the ever changing reality of existence. Most, but not all, of the knowledge contained within the pseudo-sciences, religions, and theologies are merely unproven hypothesis. However, there are also many unproven hypothesis and "Hypothetical" things within the body of Scientific knowledge, but most Scientific knowledge is more useful than not useful. Using, applying, and testing Hypothesis is how we determine right from wrong. **Don't Believe the Hype, Test it!**

3.2. The Experiments of Science

EXPERIMENTS

The "Experiments of Science" is the stage in the "Science of Sciences" where all of the guesses, assumptions, beliefs, hopes, and speculations that were generated as Hypothesis are put to the test of experimentation and are attempted to be used, applied, and practiced. The Experiments of Science, and in Life, are the **"Showdown between Right and Wrong"** information, with "Wrong Information" (Bad or Evil) failing every time and "Right Information" (Good or God) succeeding every time. The "Experiments" stage in "The Science of Sciences" is called **Armageddon** or **"Judgment Day"** in some religions. Experiments are **"Tests of Faith"** where one literally puts their faith, beliefs, and hypothesis to the test. The Experiments of Science is where thoughts and ideas (represented by the Rational world of the mind, Moon, and the dead) transform into reality (represented by the Empirical world, the Sun, and life); and thus is the point of **Creation** where abstractions become concrete. If a belief or hypothesis is based on Sound Reasoning, then it will not fail, but rather be successful when put to the test of Experimentation. Experiments are the metaphorical "rebirth of the Sun from night" and the "resurrection of the dead." We use a horizontal line on top of a triangle as our symbol for "The Experiments of Science" as this symbol resembles a scale or balance that enables the weighing or judgment of two things.

The etymology of the word "Experiment" is "to have experience" and Experimentation enables us a means to obtain Evidence to disprove our reasons, beliefs, and hypotheses. When performing an Experiment, the Experimenter must consider all the factors that contribute to the experiment, and must **Control** as many of those contributing factors as possible to accurately test the intended hypothesis or belief. When controlling the factors in an experiment, a general rule is "**Look but do not touch**," that is to say, "Experience, but do not influence, interfere, or attempt to affect the outcome of the experiment in order to learn." Controlling the factors in experimentation is considered a **Laboratory Experiment** as opposed to a **Field Experiment** where there is less control or rather **Nature is in control** of the factors in the experiment. It depends on the nature of the hypothesis being tested as to whether or not the Experiment will take place in a Human controlled Laboratory, or in Nature's controlled Laboratory called "The Field." For example, Astronomy experiments have to be Observational in the Field because as of right now, it is impossible for Humans to control Astronomical movements. However, certain experiments in Physics and Chemistry can be conducted in a Human controlled Laboratory setting. In the event where the Experimenter may desire to control certain factors that Humans cannot control, a **Simulation** or "**Virtual Reality**" may serve the purpose of meeting this aim. However, any evidence that is obtained in a Simulation of Reality must eventually be tested and experimented on in Reality to determine the validity of the conclusions and experiment.

Experimental trials often lead to an outcome where a Hypothesis that was once believed to be true, is proven to be false, which indicates error. The experimental process is often a repeated process of **Trial-and-Error** until "Right Knowledge" is obtained and the desired goal is accomplished. In addition to "Trial-and-Error," we consider Experimentation to be synonymous with **"the process of elimination,"** and **troubleshooting** as all of these methods involve the repetition of a process or experiment until the Hypothesis is supported or the desired goal is achieved. However, "insanity is repeating the exact same hypothesis over and over, and expecting different results from the experiment." When a hypothesis fails when put to the test of experimentation, the hypothesis must be revised and the experiment must be revised and repeated to account for the changes. Because we never really know the outcome of an experiment, experimentation is often performed with the attitude of **"Here goes nothing,"** as if to say, this experiment will show if my thoughts are right or wrong. Experimental methods are the bridge between **Experience** and the **Mental** realm which is why you see both the words "Experience" and "Mental" in the word **"Experi-mental."**

The reality of the diversity of Human life is that we are all different formulations or experiments to test different hypothesis, and our lives are the experimental test to determine the validity of the hypothesis which is us.

People can become emotionally attached to their hypothesis and beliefs because everyone wants to think that they "know" what is right. Since we have a brain and senses which enables us the ability to acquire information and knowledge, then it is indeed one of our purposes in existence to "know" things. However, the emotional attachment to hypothesis and beliefs biases the individual and actually hinders the process of "knowing." People say that "Fear" stands for "False Evidence Appearing Real," but fear can also be due to real evidence or real experimental evidence, which is why many people fear putting their hypothesis to the test of experimentation in the Science of Sciences. People fear the "Judgment Day or Armageddon" which is experimentation because they may find out that what they believe is wrong! Experimentation is **"The Great and Dreadful Day of the Lord"** which is great for those hypotheses that are correct and dreadful for those hypotheses that are incorrect. As we explained earlier, the emotion called "Fear" is the result of a Negative Expectation (Hypothesis) about experimental factors that the individual has no control over (Natural or Field experiments). Thus in Nature's Laboratory, only those hypothesis which positively predict Natural outcomes will be unafraid of experimentation, whereas those individuals who formulate hypothesis and beliefs that are not in tune with nature never really know what the outcome of the experiment will be, and thus feel fear. Therefore, in order to overcome fear, one's mind must generate hypothesis that are in tune with nature and that will successfully past the test of Experimentation.

3.3. The Experiences of Science

EXPERIENCE

The "Experiences of Science" is the stage in Empirical method of "The Science of Sciences" where one obtains Evidence in the form of information from the External world. The word "Empirical" means "guided by Experience," and thus "The Experiences of Science" is what guides "The Science of Sciences." This is the stage in "The Science of Sciences" that enables the practitioner to **Examine Evidence to gain Experience**. Information that is Experienced enters into the mind by way of the senses and sensory organs, thus we use an **"Eye"** as our symbol for "The Experience of Science" because the Eye is the sensory organ that is closest to our Brain.

There are basically two types of Experience: **"First Hand Experience"** and **"Second Hand Experience"**, also called **Primary Experience** and **Secondary Experience** respectively. "First Hand" or Primary experience is the Best form of Experience because information obtained by way of "First Hand" experience is **Directly** detected by the individual's mind by way of their senses. "Second Hand" or Secondary experience is information that is experience by another person and relayed, retold, or re-transmitted to someone else, and thus this metaphorically "reflected light" or "retold information" is not as intense or potent as the original Primary Experience.

Information obtained by way of "Second Hand" or Secondary experience can become more and more distorted the more times it is reiterated. However, sometimes "Second Hand" experience is the best and most reasonable way to obtain certain information. For example, it is impossible to get Primary "First Hand" experience about events in the Past that were in a different place, and therefore to obtain information about those events, it is necessary to obtain "Second Hand" experience. Colloquially, information obtained by way of "First Hand" experience is called being **"Street Smart"** whereas information obtained by way of "Second Hand" experience is called being **"Book Smart."**

The majority of information that we know has been obtained by way of Secondary Experiences. We may have had a Primary Experience with a Secondary Source, but that information is still "Second Hand" Experience. Consider your knowledge about the internal organs of the Human Body. Most people have not ever seen the inside of a Human Body, let alone their own body, but yet by way of Second Hand Experience you have knowledge and information about internal organs. You use the knowledge you have about internal organs along with Reason to conclude that you have the same internal organs, however, if you ever have a true Primary Experience with your own internal organs, it could lead to your own destruction.

Information obtained by way of Secondary Experience must be tested for validity and consistency, and Information obtained by way of Primary Experience must be tested for validity and consistency also. With Secondary Experience you test to determine if the information relayed to you was correct, and with Primary Experience you test to make sure what you Experienced was indeed reality (sanity check). If you experience some empirical evidence that you have never experienced before, it may be labeled as "nonsense" or para-normal. When an experience is encountered that may not fit into your pre-existing assumptions based on your previous experiences (either empirical or rational), you must be willing to test the experience and your understanding of the experience for accuracy as well as test your pre-existing assumptions or your sense of sense for accuracy. Remove emotion and attachment, let go, and grow.

Information obtained from both Primary and Secondary Experience can be tested for implications and Cause and Effect explanations. It is possible to Experience Evidence in the external world by way of the Eyes and other senses, and it is also possible to Experience Reason in the internal world of the "Mind's Eye." Whether you "sense" something externally, or you "make sense" (reason) of something internally, you are still sensing. Your "sense" of sense comes from your sensors that enable you to Experience your Existence and your reasoning mind which enables you to "make sense" of your Experiences.

Regardless if the information is obtained by way of Primary or Secondary Experiences, it still remains true that **Experience is the Best Teacher** and **Experience is the Only Teacher**! A person who becomes learned and wise by way of Experience is called an **Expert**. Teaching is the art of relaying information to another individual that was unknown to the individual prior to the relaying of the information in a way such that the information is completely absorbed and comprehended by the student or **"pupil"** at the intended level of comprehension. Also, teaching is the art of making or explaining the applicability, connections, and relationships between information that is already known by the pupil so that the connection and relationship between information becomes new information now know by the **pupil**. We find it interesting that the word "pupil" which is the **center of "the eye"** is synonymous with **"Student"** or "someone who experiences to learn." When the connections, relationships, and applications of and between information are discovered by the pupil without explicitly being explained or told to the pupil, then this situation is called being **"self taught"** (**self created**) and the new "information" (connection, relationship, or use) comes as a "revelation" to the pupil. The moment when "new revelation" comes as a result of being "self taught" by way of Experience is called **"Heureka"**! The word "Heureka" is a sudden, unexpected realization of information (an **Epiphany**), and also is the root of the word "Heuristic" (meaning "I have found or discovered it") which refers to experience-based techniques for optimal problem solving, learning, and discovery.

The word "Heureka" is phonetically similar to the words "**Heru Ka**", meaning "**Spirit of Heru**", and Heru was the Ancient Egyptian deity that represented the "Sun" (Empirical world) and also the "3rd Eye" (the mind's eye). When a "pupil" is becoming "self taught", meaning no physical person is relaying information or explanations to the pupil, the pupil still needs tools. These tools are used to communicate ideas for the "pupil" to experience. This is not truly being "self taught" because the pupil is still receiving instruction from an external source. A truly "self taught" pupil does not read about, or is not told about how, for example, the properties of electricity and magnetism work, but rather can observe and/or experience the phenomenon and learn and derive all of the explanations on their own simply by observing all of the phenomenon. In order to do this, the pupil must be in a "receptive" state. It is not realistic to be able to observe every phenomenon in a lifetime, thus second hand experiences are necessary, and the acquisition of information or knowledge from a second hand experience leads the pupil to a state of being "taught" by a "teacher" (Taut, Tight, Teach = solid). The word "Taught" is phonetically similar to "Taut" (or tight) referring to a knot, node, nut, and/or brain. The "Experiences of Science" are the means by which the most valuable Natural resource called "information" is acquired for the purpose of learning, life, creation, and survival.

3.4. The Theories of Science

THEORIES

The "Theories of Science" is the final stage of the Empirical "Science of Sciences" method (also called the "Scientific Method"). The "Theories of Science" represents the point where Empirical information that was gained by way of Experience, transforms into Rational information of the mind. The "Theories of Science" in the flow of mental information processing was symbolically called "Sunset" and also the point of transition from "life to death." The "Theories of Science" is the point where "Definitions" (Death-finite-ions) are established, thus it is not coincidence that the word "Define" and the word "Death" both share etymological meanings of "an ending or finale." The "Theories of Science" is the point at which information or "light" that is gain by way of experience, can be symbolically "reflected" or reiterated either by way of speaking, writing, drawing, or other forms of communication. In order to speak, one must first have obtained some information to speak about, and the stage of "speaking" is the "Theories of Science"; for this reason we use an **Exclamation mark** as the symbol for The "Theories of Science" since the root of the word "Exclamation" is to "exclaim" or "to speak out." Also, the exclamation mark symbol for the "Theories of Science" has an inverted triangle as its period to imply that "Theories" lead to further mental or rational analysis. Theory is the **beginning of speculation** and the **ending of operation**.

In philosophy, Theory is also called **Speculative Reason** or **Pure Reason** which establishes the beginning stages of **Logic**. Theories are interpretations and explanations of Experiences and are expressed by way of **Abstractions**. Theories are the result of tested Hypothesis. The mental creations called hypothesis that successfully pass experimentation manifest into the Empirical world as "Theories." Theories are created by the mind to explain, predict, retrace, model, and master Nature's experiences. Since "Theory" exists in order to "master" Nature and reality, it is no surprise that the word "**Theory**" and the word "**Theos**," which is another word for **God**, both share the same etymology. Thus, competing Scientific Theories is a "Conflict between the Gods." In Ancient Egypt, "The Theories of Science" was symbolized by the setting sun and also the deity called **Amun**. Amun appears in the Christian Bible as "**the Amen**" who was the beginning of "the creation of "**Theos**" or **Theories** in **Revelations 3:14** where it states: *"These things saith the **Amen**, the faithful and true witness, the beginning of the creation of **Theos**."* As the **final stage** of empirical information processing, the "Theories of Science" called Amun in Ancient Egypt, became the name that the monotheistic religions of Judaism, Christianity, and Islam called on to **end** their prayer ritual of speaking and getting to know "God" (nature or reality); namely **Amen** or **Āmīn**. In religious symbolism, Amun is called "the hidden one" that is proven by faith, belief, and trust (results of reason). Thus, Amun gives birth to the speculation of Theos, Theories, and Theorems that are used for instructing and teaching the mind.

The deity of "giving birth" in Ancient Egypt was **Tawaret** who is associated with the fact that Amun symbolically gave birth to Theos. Tawaret is seen at the center of the Egyptian **Dendera Zodiac** which is a depiction of the S.O.S.A.S.I.S. method in another

**The Egyptian Dendera Zodiac
S.O.S.A.S.I.S.**

form. Theory, as a form of "instruction for the mind," is also related to the Hebrew word **Torah** which means "teaching or instruction". In Greek mythology, the fact that Theory is the gateway from the Empirical to the Rational is portrayed in the deity **Theoria** who is said to be the **"Mother of the Moon."** **Theory** as the end of the Empirical process are not proven with reason, but rather supported with Evidence. **Theorem** as the end of the Rational process are proven with logic and reason, and not evidence. The "Theories of Science" is the gateway from the Empirical world of "The Science of Sciences" (also called The Scientific Method) to the Rational world of "The Science in Sciences" (also called the Mathematic Method). The next section will begin our discussion on the Rational world and Mathematic Method called "The Science in Sciences."

4.0. THE SCIENCE IN SCIENCES

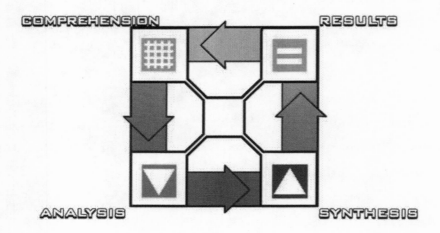

The Second part of the dichotomy that is "**The Science of Sciences and The Science in Sciences**" is the **Rational** realm of the **mind**, **mathematics**, **logic**, and **reason** called "The Science in Sciences." The word "Mathematics" means "**thought process**," thus the Methodology that is "The Science in Sciences" is similar to "**The Mathematic Method**." So, we sometimes use the terms "The Science in Sciences" interchangeably with "**The Mathematics of Sciences**." The "Science in Sciences" is symbolized by **"the square"** which represents "the finite and definite" which are distinctions and determinations established by the rational mind. The "Science in Sciences" is also associated with **the Moon** because the word **"Moon"** and the word **"Mind"** both share etymological origins.

Whereas with empiricism, you experience evidence via your senses, with rationalism, you "make sense" or "create sense" with reason and logic. Colloquially, the phrase "does that make sense" is equivalent to "is my reasoning correct?" If you experience reasoning that you have not experience before, or if you do not know how to reason soundly, then you may label certain reasoning as nonsense. Indeed, you can no more explain or discuss rational things with someone who does not know how to reason soundly no more than you can discuss or fully explain sights to a blind person or sounds to a deaf person. The "Science in Sciences" is the highest form of reasoning which is calculated by what is called **Binary Algebra** by today's mathematicians. The term "binary" refers to a number system with just 2 values: Truth and Falsehood, known and unknown, finite and infinite, male and female, Sun and Moon, etc. The science in (or Mathematics of) Religion and speculation is called Statistics and Probability by today's mathematicians. Logic is often divided into two parts, induction and deduction. **Inductive Reasoning** is drawing general conclusions and **Results** from specific examples. **Deductive Reasoning** is drawing logical conclusions and **Results** from definitions and axioms. Another form of reasoning called **"Analogy"** is useful in creating **parallels**, **metaphors**, and **examples** which enable **comprehension**. Inductive reasoning is similar to the **"Analysis of Science"** and Deductive reasoning is similar to **"The Synthesis of Science."**

In the ancient African tradition, the Goddess **Ma'at** represented Truth, Justice, Order, Balance, and Equality. "Determining Ma'at" meant trying to determine truth from falsehood or order from chaos. The affectionate term **"Ma'at-matics"** or **"Ma'at-techniques"** is used when speaking about Mathematics as the Science in Sciences because Mathematics definitely has techniques that help determine and establish Ma'at (truth, order). In order to establish order or rank, things must be assigned value, magnitude, multitude, or quantity which requires Ma'at-techniques (Mathematics). There are many fields of Mathematics just like there are many fields of Science. In terms of the "Science in Sciences" or the "Mathematics of Sciences" we are more interested in the Logic and Reasoning methodology that determines the "Science in Sciences." In a work entitled **"Egypt: Ancient History of African Philosophy,"** the writer **Theophile Obenga** outlines the method or process by which Logic or the "Science in Sciences" was processed in Ancient Egypt in seven stages:

1. tep: The stage of stating the Given Problem

2. mi djed en. Ek: This is the stage of defining all of the parts and facets of the problem

3. peter or pety: The stage of Analysis and questioning with the function of eliciting a logical predicate

4. *iret mi kheper*: The Stage of establishing a procedure, or process of showing truth by reasoning and computation

5. *rekhet. ef pw*: The stage of arrival at a clear and certain Solution

6. *seshemet, seshmet*: Examination of the Proof, Review of work, and check for generalization, confirmation, and validity

7. *gemi.ek nefer*: the stage of concluding and confirmation, Arriving at this stage means You have put fort an intellectual effort using correct, precise, and perfect procedures, and the resulting conclusion is convincing and non-contradictory

While all of the stages discussed above are present in the "Science in Sciences" methodology presented in this book, we simplify the process to the four important stages of "The Comprehension of Science," "The Analysis of Science," "The Synthesis of Science," and "The Results of Science" which are discussed in the four sub-sections of this section on the rational and reason methods of the "Science in Sciences."

4.1. The Comprehension of Science

COMPREHENSION

The "Comprehension of Science" is the first stage in "Science in Sciences" where Empirical information interacts with the Rational mind. Comprehension is the stage where empirical information, metaphorically called **"light," is absorbed** by way of experience. Since Comprehension is the ability to "catch," "grasp," or "absorb" information, we use a "Net" as our symbol for the "Comprehension of Science."

Comprehension is also synonymous with **"Understanding,"** and Comprehension is also synonymous with the affectionate terms **"Over-standing"** and **"Inner-Standing."** However, both **"Under-standing"** (meaning to stand under) and **"Over-standing"** (meaning to stand over) are just two relative perspectives of an **"Outer-Standing"** (which means to stand outside of) which must be obtained before you obtain an **"Inner-standing"** (meaning to stand inside of). You have to be "outside" of the house before you can go "inside" of the house. Collectively, all of the relative positions and perspectives of **"Under-standing," "Over-standing," "Outer-Standing,"** and **"Inner-Standing"** are included and contained in the word **"Comprehension"** (which means to completely grasp, or to encompass). Thus, we prefer to use the word "Comprehend."

The mental process of Reasoning utilized in the "Comprehension of Science" is called **"Analogy"** or **"Analogical Reasoning**." Analogy is the mental reasoning process that **transfers** information or meaning from one subject to another subject. The mental process of Reasoning called "Analogy" is responsible for creating and developing **Metaphors**, **Similes**, **Parallels**, **Examples**, and **Comparisons** between information currently being experienced and previously experienced information for the purpose of Comprehension. The Analogies, Metaphors, Similes, and Examples used in the comprehension of Science can eventually become **Abstractions**, **Mathematical Objects**, **Mythologies,** and **Parables**. The mental process of Reasoning called Analogy that is used in "The Comprehension of Science" is reasoning about the relationships of the particulars about one subject to the particulars about another subject, and is different from Deductive Reasoning (which is reasoning from the General to the specific), and Inductive Reasoning (which is reasoning from the specific to the General). The "Comprehension of Science" aids in the development of **definitions, Semantics**, and meaning of abstractions as well as the creation of new abstractions based on experienced information. The "Comprehension of Science" enables people to "speak the same language" both literally and figuratively. Because words are abstractions, Comprehension gives the individual the ability to properly mentally **Interpret** the words or information being received so that the received information **"makes sense**."

As we stated earlier, "information" has often been associated by way of metaphor and analogy to "light," therefore, since "The Comprehension of Science" is the ability to "completely grasp information," then we will discuss "The Comprehension of Science" also as a metaphor for **"absorbing light**." In Physics, a **Blackbody** is an object that is ideally able to "completely absorb light," thus we associate "The Comprehension of Science" to **The Blackbody** in Physics. Comprehension is one mind being able to show in some way that it has absorbed information transmitted to it. Comprehension is the ability to **"become one"** with that which you are attempting to comprehend while still maintaining some sense of identity. Complete experience of information enables Comprehension, whereas partial experience of information only provides perspective. **Truth** is a Measurement of how much of the information (light) was comprehended (absorbed) and when retold (or reflected) how accurate the description of the information actually reflects or describes the original information in the "Real" Natural world. If speaking in terms of "light" (information) in Optics and Physics, "Truth" is a measurement of how much the information reflected or refracted in the mind compares to the actual original incoming light (information). **The light (Information) that you have Absorbed (Comprehended) is TRUE when the "angle of incidence" (incoming empirical information) equals the "angle of Reflection" (abstraction in the mind).** In order to metaphorically "absorb or **receive** light" (literally comprehend information) you have to be in a **"receptive state."** The receptive state that enables proper comprehension is being **unbiased** for learning, and **"not leaning"** to one perspective or another. The receptive state of being unbiased is called **"Meditation"** in some religious traditions; that is, metaphorically being a **centered Plumb-bob line** on a **Triangle Level** (of course now modern builders use **Laser line Levels**). This centered, unbiased, and receptive state that enables Comprehension is also analogous to being at 90° in the **"Law of Specular Reflection"** in the **Physics of Optics**. As information (light) received from the empirical world is reflected into the mind, the degree of the information (light) is equivalent or completely comprehended when the individual is unbiased. It is said that the "Law of Reflection" in **Optics** (the science of **the Eye**) was discovered by **"Hero** of Alexandria," and this is quite coincidental since he is predated by the **"Eye of Heru"** in Egypt which was symbolic of **receiving information** as discussed in the section on **"The Noölogy of Science."**

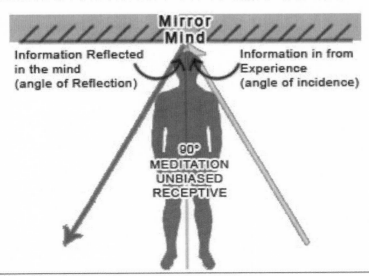

Anyone who says "They do not Judge" is not being honest, because it is one of the fundamental functions of life to receive information by way of the senses and "Judge it" or "make sense of it", and the process of judgment starts with comprehension. Extending our analogy of information to light and comprehension of information to the absorption of light (a Blackbody), then the factors that contribute to a poor comprehension would be considered **Diffraction** and **Diffusion**. Diffraction of information happens when light (information) is bent or distorted by some obstacle, making comprehension or absorption of the information more difficult and thus leading to a "mystery," a secret, problem, or unknown. Diffusion is when "light" or information is scattered which would be analogous to chaotic information or scattered information that make comprehension more difficult.

Also, **Refraction** is the "change of direction" of "light" or information, and the absorption of Refracted information would be analogous to comprehension from a single perspective (Under-standing, Over-standing, Outer-standing, or Inner-standing). The "**Refractive Index**" which determines the degree of Refraction and absorption (comprehension) of information is a quality of the "substance" (brain or individual) itself. As we stated earlier, complete absorption or comprehension of information is achieved by a **Blackbody**. Not only is a Blackbody a perfect absorber of information, but a perfect emitter (teacher or user) of information.

You know information has been comprehended when it is used or "reflected" or transmitted in the intended way it was given. A teacher can teach, and a student can appear to comprehend the information being conveyed, however, both the teacher and the student do not know if the information was Truly comprehended until the student is tested by having to either reiterate the information received or use and apply the information that was taught.

The metaphors, similes, parallels, comparisons, associations, relationships, abstractions, mathematical objects, mythologies, and parables that are created by the mind by way of Analogical Reasoning to comprehend information is necessary in order to process the information in the mind because the thoughts within the mind are conceptualizations and abstractions. It would be best if the individual could comprehend new information without having to use analogies, metaphors, and mythologies. However, the ability to comprehend new information without the use of Analogical Reasoning would be like a plant being able to **Photosynthesize sunlight** without ever getting any **water**. Even in the case of plants in the **Dessert** like cactuses that do not need much water, water is still need for growth. Just as light is symbolic of information, water is symbolic of mythology. The average Human Body is composed of about **75%** water, and therefore water is needed for life and mythology, mathematical objects, or abstractions are needed for mental life in the "Comprehension of Science."

4.2. The Analysis of Science

ANALYSIS

The "Analysis of Science" is the stage in the "Science in Sciences" (also called the Mathematics of Science) of Investigation and Inquisitiveness where information is dissected, digested, broken down, and problems are identified. The "Analysis of Science" is the "**Scimitar of Science**" and the "**Scalpel of Science**" responsible for cutting, dissecting, and breaking information down to its fundamental parts. We affectionately call the "Analysis of Science" the "Scimitar" because the Scimitar is a **sword** used for cutting, and also the word Scimitar does appear to look like **Science-Miter**-Saw. We use an inverted triangle as the symbol for "The Analysis of Science" because an inverted triangle is the geometric shape used for separation, probing, penetrating, cutting, and dissecting downward. The word "Analysis" means "to break apart" and thus is a very critical stage in "Science" which has an etymological meaning of "to separate." The Analysis of Science is the **Interrogation of Information** that enables the individual to master or "conquer" the information by "dividing" or analyzing it. The "Analysis of Science" requires the individual to be **Skeptical** and **Critical** of the information that has been received, and to investigate and probe the information with questions. The "Analysis of Science" is related to the concept called "**Scientific Reductionism**" where complex things are broken down to their basic parts.

The "Analysis of Science" makes use of the mental process of **"Inductive Reasoning"** and **"Informal Logic."** Inductive Reasoning is reasoning from some specific and particular information to something more general. Informal Logic is a type of logic used for analysis, evaluation, and criticism. Informal Logic is associated to **"Critical Thinking"** which, along with "Problem Solving" and "Divergent Thinking," is an essential component of mental **Creativity**. As we stated earlier, Creativity is associated with **"Change,"** and Change can come by way of **Construction** or **Destruction**. The "Analysis of Science" is the Destructive side of Creation. Analysis separates and "breaks down" (destroys) information using **Interrogative questions**. In English, the interrogative questions used in the "Analysis of Science" are **Who**, **What**, **Where**, **When**, **Why**, and **How**. The answers to these analysis questions will always give an answer that is one of the four aspects of nature as spoken of earlier, namely: **Space**, **Matter**, **Energy**, and **Time**. Also, it is interesting to note that the analysis question words "Who, What, Where, When, Why, and How all have the same etymological origin which is a Proto-Indo-European word spelled **HW**, which is phonetically similar to **HU**, the **Creative tone** or **"Creative Word"** of **Ptah** in ancient Egypt. Indeed, the analysis words "Who, What, Where, When, Why, and How" are **"Creative Words"** of analysis (destruction) which enable **division** for **Power** (mastery) of information for future mental building.

Answering the Analysis questions "Who, What, Where, When, Why, and How," which are six degrees of Hu - the creative word of Ptah, enables the individual to get closer to complete knowledge about an experienced phenomenon. The Analysis questions "Who, What, Where, and When" inquire into the Matter, Space, and Time of information respectively, and the answer to these analysis questions provides the individual with a basic theory and semantic definition and identification about the experienced information. The Analysis questions "Why and How" inquire into the **"Energy," cause**, or method by which some experienced information occurred. **None of the Creation-Destruction Analysis questions derived from HU can be answered with a simple "yes or no," "true or false" binary Answer.**

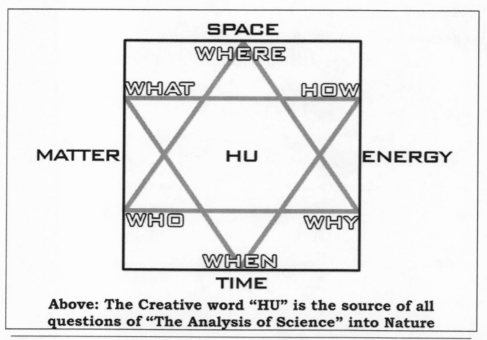

Above: The Creative word "HU" is the source of all questions of "The Analysis of Science" into Nature

4.3. The Synthesis of Science

SYNTHESIS

The "Synthesis of Science" is the stage in "The Science of Sciences" where concepts, ideas, and information come together to form new concepts and new things. The "Synthesis of Science" is the mental process that leads to an actual goal. The "Synthesis of Science" is **"putting things together"** and **"figuring things out."** Whereas "The Analysis of Science" is "Problem Finding," the "Synthesis of Science" is **Problem Shaping** which enables the solution process to begin. The "Synthesis of Science" combines information from the Rational world with Information in the Empirical world to come to a decision in the mind on what to do in reality. The "Synthesis of Science" is the mental process that puts information into "practice and use", and decides what to do, and how to act and proceed. The etymology of the word "Synthesis" means "to put together, to combine, to build, or to do." Thus, we use an upright triangle as our symbol for the "Synthesis of Science" since an upright triangle is the geometric shape associated with a continuous building upward. The "Synthesis of Science" is **Objective Thinking** using **Formal Logic** and **Deductive Reasoning**. The "Synthesis of Science" uses Deduction to Reason from general **nebulous** concepts to specific and particular ideas and actions. The "Synthesis of Science" also uses **"Practical Reason"** and **"Decision Theory"** as a means to decide the best course of action.

The "Synthesis of Science" is the mental **"Practitioner"** and **"Technician"** of Science who determines how to test, how to apply, and how to implement information. The "Synthesis of Science" includes performing **"Thought Experiments"** to mentally test how information would work in a physical experiment and in the Empirical world. The "Synthesis of Science" is also the mental powers that develop **Scientific Models**, mathematical models, and graphic models. The **"Synthesis of Science" develops the formulas that become forms**. The "Synthesis of Science" is the **Decision maker** in Science and the movement of **"will"** into **action**. Just like the "Analysis of Science," the "Synthesis of Science" involves asking questions also. However, whereas the questions asked in the "Analysis" stage are open-ended questions, the questions asked in the "Synthesis" stage are **close-ended** questions with **binary responses**. The use of binary-response close-ended questions in "The Synthesis of Science" is what leads to **decisive action: either do or do not do**. In the "Synthesis of Science" stage, rational information is reasoned about how it would function in real world scenarios, situations, conditions, and circumstances. The questions asked in the "Synthesis of Science" are conditional expressions with binary responses such as: If...then, Can, Could, Should, Would, Will, and which. Whereas the Analysis questions create by Destruction, the Synthesis questions create by **Construction**. The "Synthesis of Science" is the mental energy that shapes, forms, and moves things in the Empirical world.

4.4. The Results of Science

RESULTS

The "Results of Science" is the last stage of the "Science in Sciences" and the inevitable **Effect** of information received and processed from an experienced **Cause**. The "Results of Science" is the point where information from the Rational world of the mind manifests into the Empirical world of Reality. The "Results of Science" is the **Solution** that results from **Problem solving**, and the **Conclusion** that results from **Decision Making**. The "Results of Science" is the **Outcome** of Experiments, the **Return** of a Function, and the **Final Value** of a Calculation. We use two horizontal parallel lines called an **"Equals Sign"** as our symbol for the "Results of Science" because the "Equals Sign" is used mathematically to indicate a result, and also because the image implies duality (in this case the duality of Rationalism manifested in Empiricism). The "Results of Science" is the point where the individual **"has Created something,"** and is the **Conclusion** and **Completion** of a creative process, that will inevitably lead to more information and more creation as part of a perpetual cycle. The etymology of the word "Result" is **"to spring forward"** and is synonymous to the **"Scion** of Science."
As the point where mental abstractions become created in reality, the "Results of Science" enables the individual to **"Review"** the outcome of the Creation. In the Bible, this step is illustrated in Genesis where **_God looks at (Reviewed the Result) his creation and saw that it was Good._**

The "Results of Science" is the **"end of the Quest"** where an **Answer** is given to a **Question**. The "Results of Science" is also related to **Theorems**. Theorems are different from Theories in so far as a "Theory" would be the outcome of Empirical testing, and a "Theorem" would be the outcome of Rational testing.

The "Results of Science" or the **"Consequence of Science"** can be either positive or negative depending on your perspective. Only when the Results of Science are Negative do you **"Suffer the Consequences."** When the Results of Science are positive, you are **Happy** that the results have **Happened**. The positive "Results of Science" are **Accomplishments**, **Successes**, **Wins**, **Verifications**, **finding "Truth,"** and other positive Consequences, whereas the Negative "Results of Science" are **Failures**, **Losses**, **Accidents**, **finding "Falsehood,"** and other negative Consequences.

Ideally, one would want the "Results of Science" to produce Binding **Tautological**, Infallible Conclusions. It is best when the "Results of Science" are Conclusive, however sometimes the "Results of Science" are **Inconclusive**, **Contingent**, **Probable**, or **Plausible.** Sometimes the Results of Science end in a Paradox or Contradiction, which is indicative of a flaw in Experimentation or Reason in one of the previous stages of the S.O.S.A.S.I.S. method.

When the "Results of Science" are not Conclusive, it is likely that further investigation by way of "The Science of Sciences and The Science in Sciences" methodology will be needed. It is said that "Insanity is doing the same thing repeatedly and expecting different Results." Thus, Negative Results and Consequences or Inconclusive Results and Consequences should be reviewed and the methodology should be adjusted to obtain a positive desired Outcome. Of course, every Result is a manifestation of the Abstract into the concrete and every Result can be reviewed and experienced. Thus, the "Results of Science" leads to new Experiences, which leads to new information, which continues to move the cycle of information processing and perpetual creation in "The Science of Sciences and The Science in Sciences." The "Results of Science" is the Out-formation, which is the Result of processing and acting on Information. When we say **"Out-formation"** we are referring to anything that would be Information to someone else, but has come **Out** of you as a **Formation** and Result of your own creativity by way of the S.O.S.A.S.I.S. The "Results of Science" enables the individual to **"Reap"** the **Harvest** of the information **seeds that were sown** in the mind. The "Results of Science" is the final consequence of a sequence of mental and/or physical actions, activities, or events. With this section on "The Results of Science" we **end** our discussion on "The Science in Sciences" and move into the next section on "The Unification of Science."

5.0. THE UNIFICATION OF SCIENCE

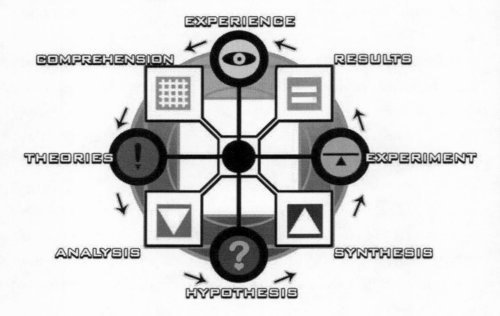

In this section entitled the "Unification of Science" we will discuss how the "Science of Sciences" works in conjunction with the "Science in Sciences." The "Unification of Science" is the complete combination of the Empirical "Science of Sciences" and the Rational "Science in Sciences" methods for information processing and life. The "Unification of Science" is therefore the complete "Science of Sciences and Science in Sciences" (S.O.S.A.S.I.S.) methodology and the picture above illustrates how the component parts of the methodology fit together and work together. Drawing a horizontal line from east to west across the above diagram separates the Empirical world at the top from the Rational world at the bottom.

Rational information is proven and processed by "The Science in Sciences" and Empirical Information is processed and proven by "The Science of Sciences." However, most information is a combination of Empirical and Rational information, and must be proved using a combination of Rational and Empirical methods which is the complete "Science of Sciences and The Science in Sciences." The S.O.S.A.S.I.S. figure shown on the previous page has eight points, and while each point has a particular area of focus and procedure, the reality is that the diagram may cycle through all eight points at each point of focus depending on the nature of the information and problem. You are able to know where you are on the diagram by the focus or outcome of your thought process or action, and therefore you gain direction in where you are going, what you should do next, or what you should revise. The Unified S.O.S.A.S.I.S. diagram is drawn to model the motion of the Sun and the Moon, and thus it cycles through positions in a counter-clockwise or east-to-west cyclic motion. The movement from point to point, or stage to stage, on the S.O.S.A.S.I.S. diagram is a movement from a Rational method (Science in Sciences) or "Thought" to an Empirical method (Science of Sciences) action. In the Unified S.O.S.A.S.I.S. diagram, Rational methods and Empirical methods alternate and cycle back and forth much like the **Binary Star System** of **Sirius A** and **Sirius B**, also called **Septet** and **Septu** in Egypt, or **Sigui Tolo** and **Po Tolo** to the Dogon of Mali.

Septet (Sirius A)
"Sigui Tolo", "the Circle",
the S.O.S.
Primary Star

Septu (Sirius B)
"Po Tolo", "the Square",
the S.I.S.
Secondary Star

The metaphorical **"Binary Star system"** that is the Unified S.O.S.A.S.I.S. method would equate "The Science of Sciences" to Sirius A and "The Science in Sciences" to Sirius B; this dichotomy is also similar to **"The Great Light** (the S.O.S.) and **the Lesser Light** (the S.I.S.)" that is spoken of in the Judeo-Christian Bible. The Unification of Science is an ongoing cycling between methods of Empiricism and Rationalism, Realism and Idealism, Objectivity and Subjectivity which are "The Science of Sciences" and "The Science in Sciences." The Quasi-empirical and Quasi-Rational method of the Unified S.O.S.A.S.I.S. process is related to the modern concepts called Logical Empiricism, **Neo-Positivism**, and Pragmatism. The "Unification of Science" is the **"Nexus of Science"** which clarifies **Nuances** and eliminates **Nuisances**.

As mentioned earlier, the concept of "Unification" in Ancient Egypt was called "Sema Tawi" which means "the union of the Two Lands" and is symbolic of a set of practices or methods that Unites the Rational mind with the Empirical world. The Unified S.O.S.A.S.I.S. method is related to the Ancient Egyptian concept of "Sema Tawi" in so far as it is a unification of Rational and Empirical methods, but also the method and the means to unite the concepts and ideas in the mind with actual physical creations in reality. The Unified S.O.S.A.S.I.S. method is **"Squaring the Circle"** and also **"Circling the Square."**

Whereas the **Empirical** world of **"The Science of Sciences"** is **"Operative"** or **"Exoteric"** (**physical**), the **Rational** world of **"The Science in Sciences"** is **"Speculative"** or **"Esoteric"** (**spiritual**). The Unified S.O.S.A.S.I.S. method is **"The Operation of Speculation and The Speculation of Operation."** In order to "Operate" or "act" you must first "speculate" or "think," however, you cannot "Speculate" or "think" without receiving information from "Operations" or "actions." Operation in the physical world is "acting" or "doing", and operation in the mental world is logic and reasoning. Speculation in the physical world is observing and experiencing without acting, and speculation in the mental realm is guessing and assuming without reasoning. However, it is the nature of the Physical to be "Active" or operative, and the nature of the Mental to "Passive" or speculative.

Science is an attempt to make Unknown information into known information (or make the infinite into the finite); that is metaphorically speaking, "square the circle." What is interesting is that "The Science in Sciences" (Mathematics) has shown that the circle can be squared in concept (a square of length $\sqrt{\pi}$ would "square the circle), but it cannot be done in practice (experimentally) with a compass and a straightedge even though both shapes have 360 degrees. This means that even though "the light" (symbolic of "The Science in Sciences", also the Square, the finite, the masculine) may think it will succeed and want to try to consume "the dark" (symbolic of "the Science of Sciences", also the Circle, the infinite, the feminine), it cannot be done as of yet. Thus, Science will always exist as long as this problem persists. As long as there are differences (duality), then there will always be conflict (or WAR) in some form, and the greater the magnitude of the differences, then the greater the magnitude of the conflict. From the point of perspective of the two sides of the dichotomy, when similarities outweigh or out number differences then that is peace. The S.O.S.A.S.I.S. is a means to Liberation, Problem Solving, and "Reconciling differences." The Unification of "The Science of Sciences" with "The Science in Sciences" is the **"Nuptials of Science"**, with the etymology of the word "Nuptials" coming from **Nupti** meaning **"to marry or unite."** The inward and outward flow from Rationalism to Empiricism is what leads to **Creation**, and thus the Unification of Science is also the **"Nymphomania of Science."**

In the "Unification of Science," (which is the complete S.O.S.A.S.I.S. method) starting from the point of **Experience**, information is received by way of the senses into the mind. The mind then uses **Comprehension** and Analogous reasoning to understand, comprehend, and "make sense" of the information that it has experienced. The Analogies, metaphors, similes, mathematical objects, mythologies, parables, words, letters, numbers, shapes, and other abstractions that were created in the mind for the purpose of comprehension then become **Theories** with which the mind can process. The mind then begins to **Analyze** the theoretical abstractions (reflected light or information in the mind) using Inductive Reasoning to see how the information can be used to the benefit of the mind. From the mental analysis or decomposition of the information, the mind then generates multiple **Hypotheses** by way of **Brainstorming** and **Divergent Thinking** about the identity of, meaning of, use of, cause of, or effect of the information and then uses Abductive Reasoning to select the most probable choice from the set of Hypotheses. The mind then begins using **Convergent Thinking** to come up with ways to **Synthesize** what is known about the Empirical world with what is hypothesized in the Rational world. This mental synthesis enables the ability to use the experienced information with the new created hypothesis by way of **Experimentation**, tests, trials, and/or attempts. After the information is tried in Experimentation, the mind determines a **Result** that is then experienced, and the cycle repeats itself.

We reiterate, the words **Liberation**, **Problem Solving**, **Change**, and **Creation** are intricately and intimately related, and the "Unification of Science", which is the complete S.O.S.A.S.I.S. methodology and **Theory**, is the key to obtaining and achieving all of these **Results** and goals. A problem by definition is the "desire" (**Hypothesis**) to want to go from one state to another state. Therefore, there are two ways to solve a problem. A problem can be solved "Externally" or a problem can be solved "Internally." To solve the problem internally would mean to eliminate the internal mental state of desire, and to solve the problem externally would be to change the physical state of being. In either case, the process involves change, and a literal or figurative **Experimentation**, test, or "war." Problems arise because of desire, and desire arises because of "self preservation." Indeed "self preservation" is the goal of all beings, however it is predicated on a sense of "self." "Self" is a creation, and at the point in time where one realizes that everything is connected, and nothing is destroyed, but only transformed, then the creation of "self" transforms into a sense of "All" where there is no desire and all problems are solved. The sense of "All" is **Homogeneity** or Unification (by way of **Synthesis**) and the sense of "self" is **Heterogeneity** or Separation (by way of **Analysis**). The "self" and the "All" are **order** and **chaos** to one another from their relative perspectives of **Comprehension**; however, the existence of this dichotomy is what leads to **Creation** and all **Experiences**, **Information**, **Knowledge**, and **Science** in **Nature**.

6.0. THE PERSONIFICATION OF SCIENCE

Note: *This allegorical story which personifies The S.O.S.A.S.I.S. should not ever be retold without the meaning of the symbolic persons, places, and things therein being fully explained, less the reason for the story will be lost and the story maybe taken literally causing further confusion.*

At the "Nature Observation Learning Laboratory" (abbreviated N.O.L.L.)" there are four Scientist and four Mathematicians who work together on an ongoing project of creation, discovery, innovation, problem solving, and development. The four Scientist and four Mathematicians at N.O.L.L. are responsible for all of the World's knowledge and all of the World's inventions. When they work together in unity, harmony, and in order, then they are very efficient and productive. However, out of order and not in unity their personalities lead to an accumulation of unsolved problems and total dysfunction.

Scientist: Tina Asya Raya (who represents "The Experiences of Science") is responsible for speaking and acting out everything that she experiences, everything that she sees, hears, smells, tastes, and touches. She is an excellent communicator and speaks every language including non-verbal languages, "body language," and facial expressions. She only reports her experiences and never reports her thoughts or her feelings.

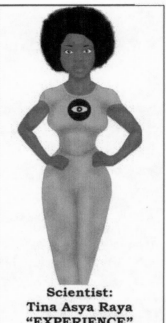

Scientist:
Tina Asya Raya
"EXPERIENCE"

Mathematician:
Sam Matemasie Anonsesye
"COMPREHENSION"

Mathematician: Sam Matemasi Anonsesye (who represents "The Comprehension of Science") is responsible for recording and capturing everything that is communicated by Scientist Tina Asya Raya. He is very observant and an excellent listener. He comprehends every form of communication and has an excellent memory. He creates metaphors, similes, and analogies which aid in comprehension.

Scientist: Ramon Kwamme Nabii (who represents "The Theories of Science") is responsible for interpreting and translating all of the information into writing that he receives from Mathematician Sam Matemasi Anonsesye. He is a very fast writer and a creative interpreter as he comes up with various signs, symbols, diagrams, and shapes to represent the information communicated to him in a written or other format.

**Scientist:
Ramon Kwamme Nabii
"THEORIES"**

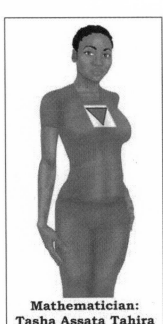

**Mathematician:
Tasha Assata Tahira
"ANALYSIS"**

Mathematician: Tasha Assata Tahira (who represents "The Analysis of Science") is responsible for analyzing and questioning all of the reports delivered to her by Scientist Ramon Kwamme Nabii. She is very inquisitive, skeptical, and very probing with her questioning. In addition to being very inquisitive, she has extremely strong inductive reasoning skills, and thus is able to infer the implications of the information reported to her. Her reports contain the answers to her questions and the inductive implications of those answers.

Scientist: Nina Naunet Kneph (who represents "The Hypothesis of Science") is responsible for receiving the reports from Mathematician Tasha Assata Tahira, and speculating, guessing, and generating hypotheses which could serve as explanations for the information reported to her. She is very intuitive, opinionated, and emotional, but she uses logic and reason to determine which of her feelings is the best possible explanation. She reports her desires and hopes about her explanatory hypothesis.

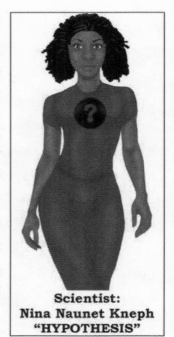

**Scientist:
Nina Naunet Kneph
"HYPOTHESIS"**

**Mathematician:
Tannon Osidan Tamir
"SYNTHESIS"**

Mathematician: Tannon Osiadan Tamir (who represents "The Synthesis of Science") is responsible for synthesizing, putting together, or building ways to experimentally test Scientist Nina Naunet Kneph's hypothesis. He has a sound sense of deductive reasoning, and is very creative. He combines his knowledge of the Rational and Empirical world to create models and methods which could accurately test Hypothesis.

Scientist: Adam Khepri Hwemudua (who represents "The Experiments of Science") is responsible for performing experiments to test hypothesis using the models and methods developed by Mathematician Tannon Osiadan Tamir. He is a skillful technician and practitioner without emotion and bias. He prefers the control of laboratory experiments but also enjoys conducting Field experiments

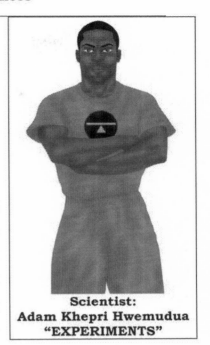

**Scientist:
Adam Khepri Hwemudua
"EXPERIMENTS"**

**Mathematician:
Tarra Nyansapo Tawaret
"RESULTS"**

Mathematician: Tarra Nyansapo Tawaret (who represents "The Results of Science") is responsible for reviewing the consequences and outcomes of the experiments performed by Scientist Adam Khepri Hwemudua and reporting the results and conclusions of the experiment to Scientist Tina Asya Raya. Her reports are very decisive, definite, and conclusive. She determines what information is right or wrong, true or false.

One day at N.O.L.L. on the first day of Autumn, the Scientist who stood on the circle called Circum-stance, and Mathematicians who stood on the square, decided that they would no longer work together. The Scientist decided they would work outside the building and the Mathematicians decided to work inside of the building. Throughout autumn and winter the Scientist continued to work and operate without thinking, and attempted to gain knowledge by trial-and error. This trial and error method used by the Scientist led to several of the scientist being injured and hurt in many experimental accidents. Likewise, throughout autumn and winter, the Mathematicians continued to think and speculate coming up with bizarre concepts that seemed to have no relevance in the real world. In the spring of that year, Patahu Oboadee, the founder of N.O.L.L. learned of the disagreement between the Scientist and Mathematicians and went to N.O.L.L. in hopes to get the two groups to settle their differences and work together as a unit as before. Upon hearing of Patahu Oboadee's arrival and listening to his words, the Scientist and Mathematicians at N.O.L.L. decided to reunite. By the summer of that year, the Scientist had discovered applications for all of the Mathematicians speculations, and the Mathematicians were able to find the logical flaws in the Scientist's trials, tests, and experiments. That summer was one of the most productive periods in Scientific, Mathematic, and Technological advancements at N.O.L.L. and the Scientists and Mathematicians decided that for the good of Humanity they should always work together in order and be united.

7.0. THE APPLICATION OF SCIENCE

With this section entitled "The Application of Science," we have arrived at the end of our "**long Quest** on a **short Path**" into "The Science of Sciences and The Science in Sciences" that is called the Scientific and Mathematic method in modern scholarly arenas, and practiced knowingly and unknowingly in many cultural and religious traditions around the world. As stated earlier, the word "Science" means "Knowledge," and knowledge is acquired by way of information, and information is the essence of life and everything existing and non-existing; for how would you know you were alive or existed without information? Information is acquired to be used and applied, and the utilization of information comes as a result of processing information by way of "The Science of Sciences and The Science in Sciences." Thus, the utilization of information is "the Application of Science." The "Application of Science" is "**the Science of Creation**" and "**the Science of Creating**." The Application of Science is **the Solution to all Problems**, and "**the Science of Liberation**." The Science of Sciences and the Science in Sciences is "**the Science of God**" and "The Application of Science" is "**the Art of being God**"; that is taking on the responsibilities of becoming a "supreme human being," capable of **Knowing** all that there is to know at a particular point in time, and being **Powerful** enough to put that knowledge into action becoming **a Creator**, **Problem Solver**, and responder to the calls of lesser beings.

The "Science of Sciences and The Science in Sciences" does not tell you "**what to think**", but rather "**HOW TO THINK**" in order to be **decisive**, **productive**, and **efficient** in all activities so that any desired results are obtained. You know "what" to think by the information that you obtain by way of your senses, and "The Science of Sciences and The Science in Sciences" enables and empowers you with a means of efficiently processing and classifying the information so that it is ready for application. The Science of Sciences does not tell the individual "**what to know**" but rather "**HOW TO KNOW**" and distinguish what information is Right Knowledge, Correct Information, and Accurate Data. The application of "the Science of Sciences and the Science in Sciences" method empowers your mind to become a **Polygraph Lie Detector** capable of determining and filtering though the Truth and Untruth within information. The ability to distinguish Truth from Falsehood which comes as a result of the "Application of Science" is of paramount importance in this "**Information Age**" where the mind is bombarded with more information than at any other point in time in recent history. All information, whether right or wrong, can be used, but only "Right Knowledge" or "Correct Information" can be used exactly as it is given. The only way wrong information can be used is to correct it. Unfortunately, the application of wrong information leads to dysfunction, disastrous, and unfavorable circumstances. The attempted application of "Wrong Knowledge" leads to many trails that end in error, and wasted time and wasted energy.

The use and application of "Wrong Knowledge" manifests as debates, discussions, arguments, etc. where there is an attempt to Correct the Wrong Information. Wrong Knowledge creates inefficient thinking in the individual who wastes time and energy thinking thoughts and ideas that are baseless and not applicable in the Real World. The application of Knowledge (or the Application of Science) enables a new experience with the Knowledge affording the individual the opportunity to Know more about what they Know or learned. With the application of Right Knowledge, the individual learns more about the implications of the information, whereas with the application of Wrong Knowledge, the individual learns that what they knew was wrong.

Intellectual Obesity is a result of acquiring too much useless information (Wrong Knowledge), or acquiring too much information that is going unused and unapplied. Just like physical obesity, if the Food, Energy, or Information that you consume is not processed properly and used properly, poor health, both mentally and physically will come as a result. The S.O.S.A.S.I.S. and the Application of Science is the way to properly processing information and apply knowledge. Perhaps one of the most important characteristic acquired by "The Application of Science" (the S.O.S.A.S.I.S. method) is **Efficiency of Thought**. Efficiency of Thought is a mind that **does not waste time and energy** thinking about non-sense, silliness, ridiculous ideas, preposterous postulates, and absurd notions. Efficiency of Thought is the result of a mind that filters out wrong information and only processes Right Knowledge and Correct Information. We display the Mathematical equation for "Efficiency of Thought" below:

Efficiency of Thought = Useful Thoughts ÷ Total Thoughts

where Total Thoughts = Useless Thoughts + Useful Thoughts

Also, as your actions are dictated by your Thoughts, "Efficiency of Thought" produces **"Productivity of Action"**. "Productivity of Action" is a body that does not waste time and energy being **idle** or performing non-productive activities, or attempting to apply information to a situation where it is not applicable. "Productivity of Action" is the energy behind Creativity, and "Productivity of Action" will never "repeat the same task consistently with no changes and expect a different outcome." We display the Mathematical equation for "Productivity of Action" below:

Productivity of Action = Useful Actions ÷ Total Actions

where Total Actions = Useless Actions + Useful Actions

Efficient energy use is using less energy to provide the same level of activity. Science has shown by way of the second law of thermodynamics that no system can ever be 100% efficient, but utilizing the S.O.S.A.S.I.S. methodology brings the mind and actions to as close to 100% efficient as possible.

Information about the creation of the universe, how a star is born, how the world began, etc is necessary to satisfy the curiosity about such things. However, unless these events can actually be witnessed or reproduced, then theories, stories, and explanations about these phenomenons contain a degree of speculation and are therefore not verified sciences but rather more guessing and assuming. Also, if an individual person is able to correctly explain the causes or effects of an event they did not witness, then what is more interesting than the fact that they accurately explained the event, is HOW were they able to accurately explain the event, and can that ability be repeated. Secondly, unless the individual will be creating a universe or creating a galaxy or solar system or atom or molecule, or unless the methodology, algorithm, and/or process of this knowledge is somehow transferable and applicable in life, then once again, this information only serves the purpose of satisfying the curiosity about these events. Until the information can be applied, it is literally useless, - meaning it cannot be used! Knowledge about how the universe is created and how solar systems and atomic systems are created formed, maintained, and governed can be used as Natural examples of systems that can serve the basis for Human systematic inventions such as Economic systems, Social Systems, Education systems, etc. Amongst black Americans over the years there has been much talk about how "The System" is the root to many of the problems and woes that are encountered. Parallels can be drawn by Studying

Natural systems to serve a basis in developing various human systems that work in accordance with Nature. You can know information (Science) whether it is correct or incorrect. What determines if information is correct is if the information agrees with Reality, Nature, and is applicable. Information that is not correct also can be used, but its only uses it to disprove it and correct it. You have to know that information is wrong in order to prove that it is not right. Science is the one thing that ALL people can agree upon regardless of their religion, philosophy, or opinions because ALL people can bear witness to the fact that Science works and is practical, applicable, and useable. We observe people of different philosophical, theological, and religious backgrounds drive a vehicle, to go to a Building, to speak into a Microphone, to have their voice heard on a Speaker, on Television, or on a Radio about why they do not like or disagree with another person's philosophy or religion. We observe people of different and various religious backgrounds and philosophical backgrounds all using the same "Technology" and "Technology" is a result of "**The Application of Science**." Thus all people, whether consciously or unconsciously, unite and bear witness to the usability, functionality, practicality, and common need for Science in the face of their own personal divisive "philosophical" or religious beliefs that do not always work or function as described in their "life manuals" called Holy Books. It is our Hypothesis that Human beings have been **separated** into various cultures, creeds, philosophies, and races so that we can **analyze** each other and ultimately learn and **know** ourselves.

Thus, it says it the Quran of the Muslim's religion: *"O Mankind, we created you from a single pair of a male and a female, and **separated** you in to tribes and nations **so that you may know each other** (not that you despise each other)."* (Al-Quran, Chapter 49, Verse 13)

Since Science is the on thing that people can agree upon regardless of their religious, philosophical, or theological beliefs, then a **Science-Based philosophy**, **Religion**, or **"Way of Life"** is the "Application of Science" that will ultimately unite the peoples of the world. As we described in the section entitled "The Preface of Science," the methodologies and practices of S.O.S.A.S.I.S. can be found in the interpretations of traditional African culture and philosophy, as well as in other cultures around the world. The "Scientific Method" and the "Mathematic Method" has become trivialized and mundane, however those methods are the operative expressions of the oldest philosophy on the planet that started with African people. And, if there is life on other planets with culture and philosophy, we would not be surprised if variations of the S.O.S.A.S.I.S. method are the foundations of extra-terrestrial or alien culture because the "Science of Sciences and the Science in Sciences" is a Universal Philosophy and the Philosophy of the Universe, and the Application of the "Science of Sciences and the Science in Sciences" is the method by which Creation occurs.

The stated purpose of Religion is to serve as a method and a means to give people **Guidance and Direction** on **How to Act** and also to give people a **Path** and a **way** to "**Get to Know God.**"   The Application of Science and the S.O.S.A.S.I.S. method is the means by which all actions occur, and therefore it is what guides all Actions and enables you to know all things knowable in Nature and existence (God).   There are Eight (8) major point of the S.O.S.A.S.I.S. method as outlined in the previous sections, and S.O.S.A.S.I.S. method is depicted as a **Squared-Circle**, **Circled-Square**, or **Octagon** with 8 points geometrically.   It is not surprising then that the largest religions in the world still use the S.O.S.A.S.I.S. geometry in their most important religious structures.   For the Christian Catholic religion, an "**eight-pointed**" star or "**squared-circle**" with a **center obelisk** sits in the middle of **St. Peter's Square** in **Vatican City** in Rome.   As **Ptah** was the deification of the S.O.S.A.S.I.S. in Africa, it is not surprising that the name **Peter** means "**Rock**" and is phonetically similar to the word "**Pater**" which means "**Father,**" and both of those words are derivative pronunciations of **Ptah**.   For the Islamic Muslims, as part of required pilgrimage (Quest), they perform an act called **Tawaf** which is **circumambulating** the **cube shaped building** called the **Kaaba** ("**Circling the Square**").   For the Jewish religion, they regard the **Temple Mount** which is the site of the **Eight-sided** building called **The Dome of the Rock** as the resting place of the Divine Presence of God.

The "Squared-Circle" at the Christian Vatican	The Octagon at the Jewish Temple Mount	Muslims Circling the Square at the Islamic Kaaba

Of course all of these religious building are esoteric and speculative interpretations and representations of the original African "Spiritual Science"; the S.O.S.A.S.I.S. These religious practices are speculative Hypothesis that are not usually Applicable and tend to fail the test of experimentation in "The Science of Sciences." As people become more and more fed up with **"getting nowhere"** religion will begin to wane as the S.O.S.A.S.I.S. begins to wax, and for African people this should signify a return to traditional African culture, philosophy, and "Spiritual Science" rooted in the S.O.S.A.S.I.S.

The back-and-forth flow between Empirical and Rational methods in the S.O.S.A.S.I.S. methodology manifests in "spirituality" and religion as Prayer and Meditation. In Religion, Prayer is likened to an "external calling" or an Empirical method, whereas Meditation is likened to an "internal speculation" or a Rational method. Of course the Application of the Empirical and Rational methods in the S.O.S.A.S.I.S. is the path that leads to results.

A "Spiritual Science" is nothing more than a Doctrine, Formula, or Methodology developed as a way and guide to live their life. If "The Science of Sciences and The Science in Sciences" is used as a "Spiritual Science" and the doctrine that people use to live their lives, then it will enable people to become more efficient and productive in obtaining any and everything needed for a higher quality standard of living. The "The Science of Sciences and The Science in Sciences" enables the individual to be able to know truth from falsehood and fact from fiction. As a spiritual science, "The Science of Sciences and The Science in Sciences" enables the individual to **know Right from Wrong**, or **Good decisions from Bad Decisions**. The "Science of Sciences and the Science in Sciences" can be applied as a **Natural Science**, **Mathematical Science**, **Social Science**, or **Spiritual Science**. The application of "The Science of Sciences and The Science in Sciences" manifests as Technological creations, New Medicines, and cured diseases. Also the application of "The Science of Sciences and The Science in Sciences" in the social sciences manifests in proper living, peaceful households, functional families, thriving **Economics**, proper **Political Science**, and proper government. The "Application of Science" provides you with knowledge about how to "**treat diseases**" and also how to **treat each other**. The Application of the S.O.S.A.S.I.S. is the "art of science" which can be used to reconcile differences in any field from **Martial Arts** to **Marital Arts**.

The Application of Science is the development of culture. A culture's religion, theology, and spirituality is what establishes the **Morals** and **Ethics** of how people should act and behave within the culture. A "moral" by definition is "proper Conduct, behavior, and action" and **Ethics** is "**the Science of Morals**." Since the S.O.S.A.S.I.S. is the means by which all actions occur, then the S.O.S.A.S.I.S. can be applied to establish a culture's Morals and Ethics. Thus, **the S.O.S.A.S.I.S. is "the Science of Ethics."** People usually think of the Scientific and Mathematic methods as only an **Instruction manual** for technological development, but the methodology also has applications in Social ways. Some of the earliest applications of Science in a Social setting are found in Africa. It is no coincidence then that the philosophers who used the S.O.S.A.S.I.S. method for **Social Engineering** of **Societal Order** entitled their written documents using the word "**Instructions**"; for example "The **Instructions of Ptah Hotep**" and "The **Instructions of Kagemni**". Instructions are the step-by-step points in the S.O.S.A.S.I.S. that are the application of Science in any setting. In conjunction with the morals and ethics established by a culture's "religion," the government of a society extends the philosophical ideals of morals and ethics into ordinances to govern, **guide**, and **direct** the people called rules and laws. Again, the S.O.S.A.S.I.S. is the methodology that guides and directs action, thus the "Science of Sciences and the Science in Sciences" can be applied to develop "**Scientific-based Governments**" and "**Scientific based laws**."

The laws that exist as established by the governments of humans are not laws! Real laws are unbreakable. These "laws" declared by governments and religion are merely suggestions enforced by the technological weaponry of the person or persons who declared them. This shows the importance of "the Application of Science" called technology. The "Application of Science" based on Right, Correct, and accurate information, verified by evidence, experience, and reason is the best and highest technology, and also enables those who apply Science in Government to declare real unbreakable laws of nature. Religious Laws and Governmental Laws are merely psychological "life technologies" developed for the purpose of controlling people. When one thinks of technology, they usually think of electronics or some other mechanical device. Technology is applied knowledge or "Applied Science" in any field. Therefore, any system, growth, creation, or invention is a form of Technology, and of course "Right Technology" is that technology that agrees and works with nature. With that said, any system, whether it is mechanical, social, economic, or political, that does not work with nature is not right technology and will surely fail the test of "application and use" called Experimentation in "The Science of Sciences". This concept is stated within one of the laws of thermodynamics. The **second law of thermodynamics** states: **"The chaos of a system which is not in tune with nature will tend to increase over time."**

To those who aspire to develop social systems, study the Nature of People and use the S.O.S.A.S.I.S. to develop a system that works in accordance to people's nature. The religious concept of "Sin" is ridiculous because it implies that the individual is more powerful than the creator in that the individual could perform an act that the creator did not want performed. If you should not do something, you could not, and would not have the ability to do it – that is a Real Law of nature – unbreakable and binding. Use the "Science of Sciences and the Science in Sciences" to develop your creations and inventions in accordance with nature, and not trying to suppress, or work against nature. A creation or invention will have the nature of whom or whatever creates it; thus you must uses the S.O.S.A.S.I.S. to get yourself in tune with nature so that you "brain-children" creations will also be in tune with nature for the betterment of the world. Applying Science and the S.O.S.A.S.I.S. method to define ethics and morals would mean that "**Immorality**" would be synonymous with "**Scientific Misconduct**." "Scientific misconduct" would be to miss-conduct or improperly perform the S.O.S.A.S.I.S. method at any point. "Scientific misconduct" is fallacious reasoning, poorly designed experiments, biased comprehension, and jumping to conclusions. Scientific misconduct is not asking questions, not investigating, and accepting unverified information, falsifying results, plagiarism, omitting or suppressing information, and making improvable unsubstantiated claims.

A form of Plagiarism in Scientific Misconduct, which is the improper Application of Science, is the modern scientific practice of naming discoveries of Natural phenomenon after the person or persons who discovers it. It is very egotistical to name "Scientific" concepts discovered in Nature after the individual person who discovered it and brought it to the attention of the population; doing so is much like **Religion and Theology** and not **Science and Theory**. If the individual's theory is ever proven wrong (the Theos is found to be a "false god"), then the proof will be met with resistance by the individual and the individual's associates who are vested emotionally in the theory. Naming Natural phenomenon after people is plagiarism of Nature, and this practice only creates further bias. However, what we do observe is that many modern Scientist who have discovered some theory have names that strangely and coincidentally correspond to the theory they discovered. For example the name "Albert" as in "Albert Einstein" means "bright" and he is best known for his theory on "light"; or the name for the sub-atomic fundamental and "base" particle called a Boson is named after Satyendra Nath Bose, with the name "Bose" being another pronunciation of "Base." If scientists who discover natural phenomenon take on the names of the natural phenomenon that they discover, this is actually paying homage to nature, and is a practice that is found throughout African cultures that have "The Science of Sciences" as their cultural philosophy and "spiritual science."

All "schools of thought" and bodies of knowledge have a purpose. The basic purpose of a "spiritual science", cosmology, or "cultural philosophy" is to set up a foundation for society and outline rules and doctrines on how to live life right. Since you always speculate before you operate (that is to say you believe before you know and apply), then the 'spiritual science" or natural philosophy sets the precedence for, and influences the culture's operative science. If the spiritual science is based on lies the science will be based on lies. If the spiritual science is based on Truth, then the science will be based on truth. The spiritual science forms the initial beliefs and hypothesis that are the beginnings of scientific research, study, and investigation. If the spiritual science allows for bias and falsehood, then naturally the science will be skewed towards trying to prove a flawed hypothesis correct in the face of contradictory evidence. If a spiritual science is based on Truths and facts, then the science that grows out of that spiritual science has the potential to be the most advanced, most progressive, and most accurate science ever developed.

The S.O.S.A.S.I.S. method is the mental and physical method of creation and creating. The **Analysis** stage is the stage of "**Problem Finding**" and "**Critical Thinking**." The **Hypothesis** stage is a point of **Brainstorming** and "**Divergent Thinking**" which is also colloquially called "**Thinking outside the box**." Hypothesis is "thinking outside the box" or the square, because it is thinking on "the circle", metaphorically speaking. In reality, you never "think outside the box" because at the very moment that you think outside of one box, you have created a new box to think inside of, thus, a flexible mind is able to move in and out of "boxes" at will. The **Synthesis** stage is a point of **Convergent Thinking** and "**Problem Shaping**", and the **Results** stage is the point at which the "**Problem is Solved**", and creations manifest. Therefore, contained with the steps of the S.O.S.A.S.I.S. path are the steps of Creation and Creativity which is the most valuable talent in existence. Anything that you can think of can be made real; this is because everything initially began as a "Thought" in the "Mind" of someone or something. It is very important to classify and identify if information exists as a Concept or a Reality. If it is a Reality, then efficient thinking and the "Application of Science" would lead the individual to Create or come up with ways to use it or explain it. If it is a Concept, then the "Application of Science" and efficient thinking would lead the individual to **come up** with ways to create it and make it real if desired. Routine practices that can be done to improve your Creativity, and make you a better "Scientist" are thinking, meditating, brainstorming, brainteasers, logic puzzles, free writing, drawing, and creating artwork. You make "ablution" or ritual purification before you "Pray" or formally state your intentions to do something in the religious world, therefore you should also make ritual purification before you "Apply Science" in the real, practical, applicable and Scientific world. The practices performed to cultivate your creativity are scientific purification. Creativity, in the form of new inventions and innovative ideas, are necessary for the **Liberation** and well being of a people.

Indeed, one must have knowledge before it can be applied. The acquisition of knowledge (science) is prerequisite to the application of knowledge (technology). Not only is knowledge necessary for technology, but "Right Knowledge" (that is knowledge or information which is correct and has been verified by evidence as experience through reason) is the most applicable knowledge and the mental foundation of technology. Knowledge that is not Right (correct) should not be applicable. In science as it exists today, there are many concepts that **"work in theory"** but **do not work in practice** or reality. Indeed these concepts are non-applicable and are no different from **religious Theories** or **"Theos"** that work in "concept" but not in reality or practice. Physical invention is the development and application of knowledge that had not been developed and applied before. Mental acquisition or awareness of Right Knowledge occurs before the application of the knowledge (the invention of technology). Innovation is the use of knowledge to improve the existing application of a particular piece of knowledge (technology). In life, male and female principle comes together to create, and the same principle and concepts apply with knowledge; the S.O.S. and the S.I.S. come together to create. When the unknown becomes known, then it can be applied to create; these are the mother and father of invention. It is the application of science that enables us to expand our range of detection and acquire new knowledge and new Science. This is the perpetual flow of science giving birth to technology and technology giving birth to new science. Knowledge cannot be applied unless it is Right Knowledge.

The only use for Wrong Knowledge is to destroy it in order to build anew. In science and pseudoscience, there are many "hypothetical" and "theoretical" particles, energies, dimensions, etc. Dark Energy, Dark Matter, Tachyon Energy, black holes, etc are all "hypothetical" and thought of for the purpose of supporting and explaining some other Theory. These are not Right Sciences based on Right Knowledge. The observer could waste and spend a lifetime attempting to observe these "hypothetical" sciences and apply these hypothetical sciences rather than spending the time usefully studying and attempting to apply sciences that are verified and exist in reality rather just in theory or as a "concept". Everyone is afraid of the Atom Bomb and no one is afraid of the tachyon bomb. However, just because the knowledge/information is not applicable does not necessarily make it false. As Right Knowledge (Right Science) gives birth to Technology, and Technology gives birth to new knowledge, it may be that newly acquired knowledge may be needed to apply old knowledge. **The absence of evidence is not necessarily evidence of absence**; that is to say, just because there is no empirical proof does not mean that empirical proof does not exist. However, until information that is proven by reason has empirical evidence, then **the "absence of evidence" relegates the information to merely a concept and not a reality**. This is where we are with the hypothetical things like Dark Matter, Dark Energy, and Black Holes; there is reason to believe these things exist in reality, but perhaps it will take new technology to obtain evidence to support the reason.

The philosophy of one's culture, religion, cosmology, morals, and ethics forms the bases of all mental assumptions, beliefs, or "Hypothesis" when moving from the speculative social sciences to the operative natural sciences. Thus, if the S.O.S.A.S.I.S. is the philosophy of one's culture, religion, morals, and ethics, then it will empower that culture to become very advanced scientifically and technologically. For African people, the S.O.S.A.S.I.S. is a part of most ancient and traditional African culture and philosophy. Thus, the return to ancient and traditional African culture and philosophy with the S.O.S.A.S.I.S. as a base should place Africa at the forefront of the development of **Emerging Technologies** such as **Concentrated Solar Power**, **Bio-Fuels**, **Hydrogen Energy**, **Wireless Energy Transfer**, **Electric Cars**, **Artificial Intelligence**, **Quantum Computing**, and **Holography** just to name a few. Using the S.O.S.A.S.I.S. as a foundation for African culture, spirituality, philosophy, and government will eventually lead to the **African Scientific Revolution** and the **African Industrial Revolution**. Historians say that the Scientific Revolution that took place in Europe began with the publication of a book, and it is our hope that the publication of this book entitled "The Science of Sciences and the Science in Sciences" as well as other books written by "African Creation Energy" will be the catalyst in the African Scientific Revolution, which will ultimately lead to the African Industrial Revolution where there will be an improved quality of life, economics, and higher earning and income for African people.

The Application of Science and the S.O.S.A.S.I.S. is the means and the method to solve all problems and obtain "Liberation." The solution to most people's problems is usually quite simple, however, most people who have accumulated multiple problems usually do not want to Act on the solution and Apply Science. Therefore, the **"Problem of Problems"** is the bigger problem of why people do not want to take Action especially if the solution is simple. Problems accumulate from laziness, ignorance, and procrastination. Often times you hear people make statements such as "I know I should not do this, but I want to do it any way" when it comes to performing actions that they know are detrimental such as smoking, drinking alcoholic beverages, using narcotics, eating bad foods, and having excessive sex. A thought process in the "hypothesis" stage of the S.O.S.A.S.I.S. that leads to detrimental actions with negative consequences is a blatant disregard for the S.O.S.A.S.I.S. method of creation, and will ultimately lead to destruction. **Eternal Life is perpetual problem solving, and death is the accumulation of problems unsolved**. Applying Science and the S.O.S.A.S.I.S. will lead to the solution of all problems, including the problem of death, which will ultimately means eternal life for the practitioner, and death for those who do not follow the path of the S.O.S.A.S.I.S. Applying Science and the S.O.S.A.S.I.S. will also lead to solutions for those who have already died, and thus the S.O.S.A.S.I.S. is the path that leads to Reincarnation, or "coming back to life".

The "Science of Sciences and the Science in Sciences" is the method, the means, and the program of life, creation, education, and learning. It is said that "Education is the discovery of our ignorance," and this is very true, because through education we learn and soon discover what we are ignorant about and don't know. The "Known-Unknown" is the most spellbinding form of information if the individual does not have a proper means of transforming the unknown to the known. The **"Science of a Spell"** is to cast, or relay information that creates curiosity and "Known-Unknowns," however, **"the Science of Breaking a Spell"** is the S.O.S.A.S.I.S. method and the application of Science. There is an old saying that states "Give a man a fish and he will eat for a day, but Teach a man how to fish and he will eat for a lifetime"; the Application of Science is what teaches you how to metaphorically "fish" and eat for a lifetime and beyond. As the mental Creator, Programmer, and Controller of all of Creation, Existence, and nature, the S.O.S.A.S.I.S. is the "Program" or **"Spirit Being"** that is within all. The use of the S.O.S.A.S.I.S. method and the Application of Science brings us **peace of mind**, **clarity of mind**, and **serenity**, so we close out this section about the "Application of Science" with the **"Serenity of Science"** supplication:

Let us use our minds to discover the
Science to accept things we cannot change,
Technology to change the things we can, and the
Mathematics to know the difference

8.0. THE AUTHOR OF SCIENCE: African Creation Energy

The word **"Physics"** is a Greek word meaning **"Nature"** and has been used to refer to the scientific study of Space, Matter, and Time. In the field of Science called Physics, **"Energy"** is defined as simply "The Ability to do **work**" or the amount of **Work** that can be done by a **Force**. In the field of Science called Physics, a **"Force"** is defined as any thing that causes the **Change of Position** and hence suggests **Movement**. Only the field of Science called Physics (Nature) in the area of **Thermodynamics** (which studies the movement and transformation of Energy) can there be found any Scientific Laws (that which is not assumed but rather <u>known</u> and accepted as an undisputable fact). One of the Laws of Physics (Nature) is called **"Conservation of Energy"**. The Conservation of Energy Law of Nature (Physics) states that **Energy cannot be created** in the sense that it comes from Nowhere and Nothing into existence, and **Energy cannot be destroyed** in the sense that it ever ceases to exist, but rather **Energy can Transform or change from one form to another**. Energy has many forms including Potential Energy, Kinetic Energy, Atomic Energy, Heat Energy, Electromagnetic Energy, Chemical Energy, Gravitational Energy, Sound Energy, and all phases of Matter. Everything in existence is made up of some form of energy. In Nature (Physics), **Power** is scientifically defined as the rate or amount of time it takes for **work** to be done or **energy** to be converted. In Nature (Physics), **Radiation** is scientifically defined as any process by which the energy emitted by one object travels through a medium and is eventually absorbed by another object. In the field of Science called Physics (Nature), a **Black Body** is scientifically defined as an ideal object that absorbs all electromagnetic radiation (Light Energy) that it receives.

Creation refers to the act or process of causing something new or novel (nova) to exist or come into being. As it follows from the discussion about Energy, since everything in Existence is composed of Energy, and Energy can only change from one form to another, then Creation and Creating does not mean bringing about something new into existence from Nothing and Nowhere, but rather Transforming or changing the form of one thing, or combining the forms of several things, into the desired new thing. Creation of anything in any form is a gradual process of **Growth** and **Change** over time in which a metaphorical or literal seed, nut, kernel, or node grows and transforms into another form, figure or structure. A new creation is always the initiation or beginning of one thing and the termination or ending of another thing. Change, Growth and Decay are mathematical concepts which indicate the increase or decrease of a quantity, magnitude, or multitude over time respectively. Creativity is the level or degree of creative mental ability, and a creation always exists as a thought, idea, or concept in the mind of its creator before manifesting in the physical world. Creation occurs through Creativity, and one of the most important Creativity techniques is "Problem Solving". Problem Solving occurs when it is desired to go from one state to another state (change of position), and therefore from the scientific definitions given, "Problem Solving" and **Creativity** are both a mental **Force** which can be measured as **Energy**.

From the scientific definitions given, it follows that **African Creation Energy** can scientifically be defined as the Work, Effort, Endeavors, and Activities of Black or African people that cause a movement or change. African Creation Energy is The Energy, Power, and Force that created African people and that African people in turn use to Create. Since African people are the Original people on the planet Earth, it follows from thermodynamics that the Creation Energy of African people is the closest creation Energy of all the people on the Planet to the **Original Creative Energies** that created the Planets, stars, and the Universe. **African Creation Energy** is **Black Power** in the scientific sense of the word "Power", and this book **Radiates** African Creation Energy to be absorbed by the **Black Body**.

Amongst the **Yoruba** people in **Nigeria**, **African Creation Energy** is called "**ASHE**" which means "the power to make things happen" and the various forms of African Creation Energy are personified in the Yoruba "Orisha" (Ori-Ashe) deities. Amongst the **Akan** people in Ghana in the Twi language, **African Creation Energy** is called "**TUMI**" – the web of energy and power that exists throughout space and all of creation which was woven and designed by the Akan deity of wisdom Ananse Kokuroko. In the Congo, **African Creation Energy** is called **"DIKENGA"** which refers to the energy of the universe and the force of all existence and creation, and the thermodynamic process of the transformation of energy is depicted in the Congo cosmogram called the "Yowa". Amongst the **Mande** people of West Africa, the Creative Force called **Nyama** or **Amma** is used by the Blacksmiths as a means to forge technology for the well-being of the entire village. In Ancient Africa, Egypt, Nubia, KMT, Kush, Tamare, *etc. et al*, **African Creation Energy** was called **Sekhem** which was energy that individuals used to control the elements of nature to create anything desired. In books written by Afroo Oonoo, the African Original Creation Powers of The Universes are called **NoopooH**, and in books written by Amunnub-Reakh-Ptah, the original creative force of African people is called **Nuwaupu**.

It is the view in the Western world that *"Those who can, DO; and those who CANNOT, teach"*. This statement is used to imply that people who are able to Use knowledge once it is acquired put it into practice and those who do not know how to use the knowledge or do not have the ability to use the knowledge can only "Teach" or relay the knowledge and information on to someone else. This western view about knowledge, information, and Teachers motivates western governments to pay Teachers the lowest salaries and in turn, forces individuals who do have the "Know How" to put Knowledge into practice, out of the classroom; which only further mentally cripples the future generations. The etymology of the word "Teach" is "to speak, to tell, or to say," and the etymological sense of the word "Technology" is "to build or construct by uttering reasonable speech"; this shows a close relationship between Teaching, Technology, and "speech" (HU the creative "tone" of PTAH). Whereas Teaching is "speaking" to Instruct, Technology is "speaking" to construct. **African Creation Energy** is about **CONSTRUCTING** (DOING, TECH) and **INSTRUCTING** (TEACH) – doing what is necessary to apply all knowledge (Technology) acquirable and practical for the well being of African people NOW, and teaching, instructing, and relaying the Knowledge and the Know How to ensure the well being of African people LATER. With that being stated, the physical writer of this book - who is African by blood and lineage, a descendant of the Balanta and Djola tribes in present day Guinea-Bissau (Ghana-Bassa) in West Africa, a descendant of the Ancient Napatan, Meroe, Kushite Empire, and a Scientist, Engineer, Mathematician, Problem Solver, Analyst, Synthesizer, Artist, Craftsman, and Technologist by education, profession, and nature - has set out to develop, engineer, invent, formulate, build, construct, and create several Technologies (Applications of Knowledge) for the well being of African people world wide and has attempted to relay and teach the information that motivated and inspired the development of those technologies in a three part introductory educational series which collectively have been entitled "The

African Liberation Science, Math, and Technology Project" **(The African Liberation S.M.A.T. Project)**. The three books that are part of African creation Energy's "African Liberation S.M.A.T. project" are:

1. **SCIENCE:** (Knowledge/Information)
   The SCIENCE of Sciences, and The SCIENCE in Sciences

2. **MATHEMATICS:** (Understanding/Comprehension)
   Supreme Mathematic African Ma'at Magic

3. **TECHNOLOGY:** (Wisdom/Application)
   P.T.A.H. Technology: Engineering Applications of African Science

The purpose of African Creation Energy's "African Liberation S.M.A.T. project" is to free the minds, energies, and bodies of African people from mental captivity and physical reliance and dependence on inventions and technologies that were not developed or created by, of, and for African people. In the western world (America and Europe) there are certain technologies such as indoor plumbing, electricity, computers, internet, heating and air conditioning, television, radio, and automobiles that have become so commonplace in the lives of people that it is difficult for some people to imagine life without these technologies. Conversations with the most "Afro-Centric" or "conscious" Black person in the Western World reveal that one of the primary reasons why they prefer to live in the Western World rather than in Africa is for the technologies that they have become used to in America or Europe. A significant amount, if not all, of the inventions needed for survival and well being that African people in Africa and the Diaspora use globally have been invented and developed by individuals other than Africans. Therefore, it is observed that the majority of Creative Energies of African people in Africa and the Diaspora are directed toward purely aesthetic and entertainment creative expressions and endeavors such as music, art, and fashion. The creativity of Black or African people has dominated and influenced the world in the areas of art, entertainment, culture, and fashion, and the degree and level of African Creation Energy can be seen and examined in the degree of Creativity that is exhibited by African people in the areas of music, art, culture, entertainment, and fashion globally. However, Creative expressions such as music, art, and fashion solely for the purpose of aesthetics and entertainment should only come after Creative Energy is used for the purpose of creating, developing and inventing everything needed for survival and well being. Therefore, the solution to Liberate African people from the spellbinding reliance and dependency on Western, European, or Non-African technology is to engineer, invent, and Create African technology. The word "Technology" means "applied knowledge", therefore, to be dependant on the "Technology" of another is to be dependant on someone else to think for you! Technology or "applied knowledge" in the broadest sense is the entity that is the most spellbinding and captivating for African people by non-Africans. Religion is a technology or invention that applies amongst other sciences, the sciences of Psychology and Sociology which spellbinds and captivates the minds of African people. Money is a technology or invention that applies amongst other sciences, the Science of Economics, and spellbinds and removes African people away from real Natural resources. Technology in the form of Inventions applies science to make life easier and spellbinds the Minds, bodies, and energies of African people making African people dependant on the thinking and creativity of other people. Religion, Money, and Technology are the three most spellbinding and captivating forces that African people must be Liberated from in order to advance, and all of these entities fall under the umbrella of "Technology" (Applied Knowledge). Therefore, it is the goal of African Creation Energy to Liberate African people from the "Spell of Technology".

It was Technology, either in the form of weapons, doctrines, or other inventions, that enabled Europeans to colonize, enslave, conquer, perpetuate, and enforce their rule over African people and the world. Anything that threatens the survival, well-being, and free expression of African culture and concepts is indeed oppressive and requisite for Liberation. Liberation from oppressive entities may require physical conflict, but until African people create African technology that is competitive with, or better than those of our oppressors, engaging in a physical campaign for liberation is understandable from an emotional standpoint, but premature in terms of being sincerely practical resulting in a positive outcome. It is unintelligent to go up against tanks and machine guns with literal or metaphorical sticks and rocks, this action will only lead to slaughter. It is likewise unintelligent to go to war using weapons and technology that was invented by the very foe that is being fought. The metaphorical SPEAR AND SHIELD (offensive and defensive) technology of Africans must be advanced, upgraded, and progressed to ensure the salvation of African culture and well being after Liberation. The Asian writer Sun Tzu wrote about "The Art of War", however, having the ability to create the necessary technology to ensure that African people are not colonized, enslaved, exploited, or taken advantage of, and ensuring that traditional African culture is preserved is "The Science of War". Technology in the forms of weapons (such as guns, bombs, missiles, lasers, etc.) that threaten the survival, existence, and well-being of Africans are the fear inducing, supporting, and driving forces of the success of other technologies such as Law systems, Religious systems, and economic systems because without the "enforcer" technologies in the form of weapons, African people would immediately see the flaws in the Law, Religious, and Economic systems and would quickly abandon them. In addition to weapon technology as the physical active "Enforcer" of the spell-binding and captivating mental technologies such as Law, Religious, and Economic systems, survival technologies that make day-to-day life easier and simpler (such as electricity, plumbing, automobiles, phones, etc.) are the dependency creating passive enforcers and active encouragers of African people to willingly partake and participate in non-African

Culture, Law, Religious, and Economic systems. Furthermore, the misunderstanding about the inner workings of technology developed by someone who you have let do your thinking and creating for you can and eventually will lead to the same wild speculations and misunderstanding about how nature worked, functioned, and operated, thus created "Technological Spook-ism". "Technological Spook-ism" are the unreasonable and irrational speculations performed by people who do not have the knowledge or ability to develop technology and do not know how technology works, or its capabilities, thus create "Technology Myths" to explain it.

The importance of Liberating African people from the dependency and reliance on Non-African technology is because an invention or creation is just like an offspring; it comes from its creator and therefore has its creator's Nature and likeness. A "Brainchild" in the form of a concept, idea, invention, or creation is just like an actual child in that certain traits and characteristics of the creator are passed on to the creation. Therefore, a person, place, or thing has the nature and likeness of whoever created it. If the creator or inventor of a piece of technology needed for survival has a Natural disposition that is not in favor of the survival and well-being of African people, then that piece of survival technology is actually detrimental to the survival and well-being of African people. In the western world there have been many Scientist, inventors, engineers, doctors, etc of African descent that are responsible for many technological advances in the western world. Lewis Latimer, George Washington Carver, Benjamin Banneker, Elijah McCoy, Garrett Morgan, Granville T. Woods, Philip Emeagwali, Patricia Bath, Marie Van Brittan Brown, Jane Cooke Wright, Louis Tompkins Wright, etc. and other members of organizations like the "**National Society of Black Engineers** (NSBE)" were African people and organizations that contributed significantly to the revolutionary, technological, and industrial advances that made the United States into the world power that it is today. In order for Africa to advance significantly in the same manner, African people worldwide must contribute their respective creative energies to the development and advancement of Africa. There have been many Afro-centric (African centered) revolutionary **movements** (forces or energies) and organizations for a variety of causes that have emerged over the years. Economic movements such as "Black Wall-street," "Ujamaa - Cooperative Economics" and others have been aimed at empowering African people financially and economically. Movements such as "The Black Panther Party", "Pan-Africanism", "Uhuru", and "African Socialism" have been focused on empowering African people politically; and numerous organizations have been developed to promote, teach, and empower African people spiritually, religiously, historically, and culturally. The African Creation Energy **Movement** (or Force) is one that not only **empowers** the **Black Body** of African people creatively, scientifically and technologically, but also Creates **Power** from African Science and Technology and seeks to be the catalyst in the **African Technological and Industrial Revolution**. If African people need something, then we should possess the Creativity, Knowledge, and Will to create it; in Africa this is referred to as *"**Call and Response**"* – the basis of problem solving and a fundamental aspect of Creativity. As long as African people rely on others to do our Thinking for us is as long as the problems of African people will persist. African Creation Energy provides the creative insight and motivation as well as the correct scientific knowledge to respond to the calls, answer the questions, solve the problems, and create solutions for African people.

The phrase "Reinventing the wheel" is a metaphorical phrase used to mean the duplication, re-constructing, or re-inventing of a basic idea, object, method, technology, or other invention that is known to already exist and be in use. The metaphor "Reinventing the wheel" is based on the fact that the wheel is a prototype, basic, simple, and fundamental human invention that is the corner stone of other technologies and not known to have any operational flaws. Therefore, the metaphor "Reinventing the wheel" attempts to symbolically express a creative endeavor that would seem pointless and add no value to the object, and would be a waste of time, diverting the inventor's energy from possibly more important problems which his or her energy could be put to use. However, African Creation Energy reiterates, "A creation or invention (person, place, or thing) has the Nature of whomever or whatever created or invented it". Therefore, if the metaphorical wheel was created or invented by someone or something who's

Nature is not in favor of the survival and well-being of us African people, then us African people will be, and are using tools and inventions of our own destruction. For this reason, African Creation Energy must Re-invent the metaphorical wheel whenever necessary. Of course, any duplication, reinvention, or reproduction of a creation or invention that already exists must be done within Reason and in accordance with Nature for the survival and well being of African people everywhere.

At this point it is reiterated, the word "Technology" means "Applied Knowledge" and therefore does not just refer to electronics, tools, gadgets, and/or gizmos; but rather to any and all persons, places, and things that can be created, developed, formulated, manufactured, engineered, or invented by applying and using knowledge. This book is written by African Creation Energy for African Creation Energy to develop not just "Free Energy" in the sense that you can light your house and drive your car for free, or no continuous monetary cost, but "**Liberation Energy**" in the sense that you are mentally liberated to know how to do what ever is needed to accomplish any desired goal or create any solution. African Creation Energy is Real Black Power, not just a cliché, but Creative Black Power that can light your house, run your car, teach your children, and guide your government.

"The Ethiopian Race is not only physically captive to MANKIND but also captive to his CULTURES and INVENTIONS, and this means Wooly-Haired People are mental captives to adverse forces as well as physical. Mental captivity is the worst kind of captivity because it means that the captive is mentally dependent on the captor. Mankind is all peoples with straight hair by Nature… If we need something we must work and get it or **create it**. *So long as we remain mentally dependent on other races to do our thinking and plan our progress, that is how long we will remain in the GHETTOES and SLUMS and the POVERTY and DISEASES that they produce…The REASONABLE and PRACTICAL worldwide solution is MENTAL LIBERATION which produces* **ETHIOPIAN CREATIVENESS** *(the ability to create everything we need) and ETHIOPIAN INDEPENDENCE (the ability to think independent and act independently of other races)… A person, place, or thing is alien to us Ethiopians when it is not of and for us by Nature. A person, place, or thing is in the Nature of whoever created it."*
-The Nine Ball Liberation Information

"The world is indebted to us for the benefits of civilization. **They stole our arts and sciences from Africa.** *Then why should we be ashamed of ourselves? Their modern improvements are but duplicates of a grander civilization that we reflected thousands of years ago, without the advantage of what is buried and still hidden, to be reflected and resurrected by our generation and our posterity… There is no height to which you cannot climb without the active intelligence of your own mind. Mind creates, and as much as we desire in nature we can have through the creation of our own minds…in your homes and everywhere possible, YOU MUST TEACH THE HIGHER DEVELOPMENT OF SCIENCE TO YOUR CHILDREN; AND MAKE SURE THAT WE HAVE A RACE OF* **SCIENTISTS PAR EXCELLENCE.** *For in religion and science lies our only hope to withstand the evil designs of modern materialism"*
~Marcus Mosiah Garvey

Daoud of Ptah
Balanta-Bassa Ajamatu
Osiadan Borebore Oboadee
Baba Phut Hotep
Kwadwo Padɔ

0.0. THE AFTERWORD OF SCIENCE by Tebnu Akhir

Throughout history there has always been a conflict between two opposing forces. This historical reality does not only exist within mythology as "brothers fighting brothers", but it also exists within the realm of Science. In Science, this concept of opposing entities is seen for example as the positron (or anti-matter particle) clashing with the electron (or matter particle). Scientifically, using reason, one can come to the conclusion that in order for there to be any conflict, there must be an equal distance for opposition to occur; this fact is prevalent in Science and Mythology where things or people must share a "common ground" or idea, and it is the "difference of view" that sets them apart for conflict or opposition. Such a reality can be called "Equal Opposition" in which things or people are equal in Nature, but opposite in degree (hot and cold), which results in conflict. Since it is in the nature of people to clash because of "Equal Opposition", then the different ideologies that people construct can also inherit the same equal opposition. Evidence of conflicting ideologies based on "Equal Opposition" can clearly be seen in the concepts of Atheism and Religion; for the two are but opposite extremes of each other. Atheists are usually Scientific minded people, and try to disprove the existence of God using Science. On the other extreme, Religious individuals think Science and Religion are opposites, and should be separate. However, Atheism validates the existence of God or "Theos" within the very word "Atheist."

The etymology of the word "Atheist" is "against God" or "against Theos". How could one claim to be against a God that to them does not exist? Thus, the word "Atheist" validates the existence of God or "Theos" rather than disproves it. According to the "Scientific Method", a "Theory" cannot be proven, but rather only supported or refuted by evidence. Thus, as the words "God" and "Theory" are related in the Greek language by the word Theos, Atheism will never disprove the existence of God, and likewise the same holds true for Religion. Religious people seem to base the existence of God on their own existence. In the most obvious case, religious people base God's existence on their own material wealth. In the same manner that Religion says one cannot see God, Science says one cannot see the electron. The "Electron" in Science is "Theoretical" whereas God in Religion is Theos. Yet, the manufacturing of all electrical equipment is based on the "Electron Theory" in the same way that Religion accredits God with placing all the stars in the sky. Thus, the common ground for both of these extremes is Reality; a Reality that first begins with you (**I**) the observer attempting to explain the phenomena of worldly events. So, both of the mental concepts of Atheism and Religion are interdependent just as "evil" depends on "good". The Tao Te Ching articulates this concept as "to speak of light is to bring in darkness, and to speak of good is to introduced bad." In Ancient Egypt, both Science and spirituality were intricately woven into the fabric of society. For example, in Ancient Egypt Ptah can be seen as a master craftsman and a divine representation of the unseen God

Amun. Thus, one who profits from ripping the fabric of Science and Religion will forever try to point out the differences between the two. Yet, when one adopts both Science and Spirituality, a dynamic Balance can be obtained as society will reap the benefits of a "Spiritual Science". "The Science of Sciences and The Science in Sciences" portrays eloquently the dynamic balance of Science as a form of "Spirituality" or "way of life". I met the author of "The Science of Sciences and The Science in Sciences" while he was working on a projected he called "African Liberation Science, Math, and Technology", and it was apparent from our conversation that establishing that Balance one of his goals of the project. I wish success to the author in his endeavors, and I am proud to have been able to present the afterword for a great Scientific work of craftsmanship.   *~Tebnu Akhir*

REFERENCES and RESOURCES:

- "African Cosmology of the Bantu Kongo" by Dr. K. Bunseki Fu-Kiau
- "African Origins of Civilization, Religion, Yoga Mysticism and Ethics Philosophy" by Muata Ashby and Karen Ashby
- "Blacks in Science: Ancient and Modern" By Ivan Van Sertima
- "Black Pioneers in Science and Invention" By Louis Haber
- "Egyptian Yoga: The Philosophy of Enlightenment" by Muata Ashby
- "Egypt: Ancient History of African Philosophy" by Theophile Obenga
- "Introduction to The Nature of Nature books 1 and 2" by Afroo Oonoo
- "Self Healing Power & Therapy: Old teachings From Africa" by Dr. Kimbwadende Kia Bunseki Fu-Kiau
- "Stolen Legacy: Greek Philosophy is Stolen Egyptian Philosophy" by G.M. James
- "The Ancient Egyptian Buddha: The Ancient Egyptian Origins of Buddhism" by Muata Ashby
- "The Memphite Theology" www.kheper.net/topics/Egypt/Memphis.html
- "The "Memphite Theology" http://www.maat.sofiatopia.org
- "The Maxims of Ptahhotep" http://www.maat.sofiatopia.org
- "The Origin of The Buddha Image & Elements of Buddhist Iconography" by A.K. Coomaraswany
- "The Nine Ball Liberation Information Count I, II, III, IV" Written by Wu Nupu, Asu Nupu, and Naba Nupu
- "The Science of Sciences" by Herbert Benson and Chinmaya Publications
- "What is Nuwaupu?" by Amun nebu Re Akh Tar
- "Wu-Nuwaupu?" by Amunnub-Reakh-Ptah, Paa Hanutu: Djedi

PHOTO CREDITS:

Page 20, The Dharma Wheel, User Esteban.barahona, Author Shazz, from Wikipedia

Page 21, Bagua - Fuxi earlier heaven bagua, Date: 2010-07-09 15:23 (UTC), User: Machine Elf 1735, from Wikipedia

Page 24, The Dome of The Rock mosque, in the Temple Mount, Date 3 January 2010, Author: David Baum, from Wikipedia

Page 25, Shot of The Octagon from UFC 74 ; Clay Guida vs. Marcus Aurelio. Clay Guida taking on Marcus Aurelio, Date: 05:40, 6 September 2007, Author: Lee Brimelow

Page 27, Sri-Yantra Mandala, Date: 10 December 2006, Author: N.Manytchkine, from Wikipedia

Page 35, The Shabaka Stone at the British Museum, Date: 14 September 2007, Author: Markh at en.wikipedia, from Wikipedia

Page 38, Image of Ptahhotep from http://www.ancient-egypt.co.uk/people/the-Art.htm

Page 43, 4 Baboons around the "Lake of Fire", Papyrus of Ani, from http://www.bibleorigins.net/hellsorigins.html

Page 53, The deity Hapy performing Sema-Tawy, Date: 06:17, 11 January 2008 (UTC), Author: Jeff Dahl, from Wikipedia

Page 59, Triangle Level of Sennedjem from the tomb of Sennedjem, The Egyptian Museum in Cairo, http://www.nga.gov/exhibitions/2002/egypt/ts_level_full.htm

Page 63, Unknown Nubian Egyptian Scribe Statue, from http://web.ukonline.co.uk/gavin.egypt/the.htm

Page 63, Babalawo performing Ifa, from http://www.uflib.ufl.edu/afa/reserves/poynor/arh3525/arh3525-3.html

Page 93, An overview of particle physics, Date: 6 January 2009, Author: Headbomb, from Wikipedia

Page 185, Aerial view of the Temple Mount, Source: http://www.bu.edu/mzank/Michael_Zank/Jerusalem/aerial.jpg

Page 185, Aerial view of St. Peter's Square in the Vatican, Source: http://www.canadafirst.net/our_heritage/solstice/sol-stpeters.jpg

Page 185, Muslims Circumambulating the Kaaba, Source: http://www.hamzaqulatein.co.uk/files/gimgs/24_23930469006b40ba2274.jpg

INDEX

Made in the USA
Lexington, KY
07 January 2013